Integrated Infrastructure for Sustainable Improvement of Movement and Safety in Urban Road Corridors
The Case of Dar es Salaam, Tanzania

T0186430

Integrated Infrastructure for Sustainable Improvement of Movement and Safety in Urban Road Corridors

The Case of Dar es Salaam, Tanzania

Een geïntegreerde infrastructuur voor duurzame verbetering van mobiliteit en veiligheid op stedelijke verbindingswegen
Een casestudy van Dar es Salaam, Tanzania

Thesis

to obtain the degree of Doctor from the
Erasmus University Rotterdam
by command of the
rector magnificus

Professor dr H.G. Schmidt

and in accordance with the decision of the Doctorate Board

The public defence shall be held on
April 26, 2012 at 16:00 hrs
by

Christina Kayoza
Born in Bukoba, Tanzania

ISS International Institute of Social Studies

Doctoral Committee

Promotor
Prof.dr. M.P. van Dijk

Other members
Prof.dr.ir. A.E. Mynett, UNESCO-IHE
Prof.dr. L. van den Berg
Prof.dr. W.A. Hafkamp

CRC Press/Balkema is an imprint of the Taylor & Francis Group, an informa business

Published by:
CRC Press/Balkema
PO Box 11320, 2301 EH Leiden, The Netherlands
e-mail: Pub.NL@taylorandfrancis.com
www.crcpress.com - www.taylorandfrancis.com

ISBN 978-0-415-62714-6 (Taylor & Francis Group)

DEDICATION

I dedicate this Thesis to Jehovah God Almighty Creator of Heaven and Earth who enabled as well as gave me the energy to accomplish this Thesis.

TABLE OF CONTENTS

DEDICATION..IV
ACKNOWLEDGEMENT XVIII
ABSTRACT...XIX
LIST OF FIGURES..IX
LIST OF TABLES XI

PART ONE: INTRODUCTION, THEORETICAL FRAMEWORK AND

METHODOLOGY 1

1.0   INTRODUCTION 2

1.1   BACKGROUND TO THE RESEARCH 2
1.2   MOVEMENT 2
1.3   SAFETY OF MOVEMENT..5
1.4   STATEMENT OF THE PROBLEM 7
   1.4.1   Research Objectives...7
   1.4.2   Research Questions 8
   1.4.3   Research Hypotheses 8
1.5   STRUCTURE OF THE THESIS 8

2.0   INTERACTION AMONG INFRASTRUCTURE COMPONENTS IN URBAN ROAD CORRIDORS:
CURRENT STATE OF KNOWLEDGE 10

2.1   INFRASTRUCTURE COMPONENTS IN URBAN ROAD CORRIDORS.......................................10
2.2   ROADSIDE ZONE 11
   2.2.1   Edge Zone...11
   2.2.2   Furnishings Zone 11
   2.2.3   Throughway Zone 12
   2.2.4   Frontage Zone..12
   2.2.5   Travel Way 12
2.3   ANALYSIS OF INTERACTION OF URBAN INFRASTRUCTURE 13
2.4   EMPIRICAL AND ANECDOTAL EVIDENCE OF INTERACTIONS ON ROAD CORRIDORS 14
   2.4.1   Structural Performance Relationship between Pipes and Road Pavement.......14
   2.4.2   Buried Pipe Performance Factors in Structural Performance of Pavements....20
   2.4.3   Effects of Replacement and Maintenance of Buried Utilities and Road Pavements.................21
   2.4.4   Excavation Damage 28
   2.4.5   Functional Interactions 29
2.5   IMPLICATIONS OF NATURE OF INTERACTIONS...31
2.6   CURRENT STATE OF PRACTICE 32
   2.6.1   Development and Management of Infrastructure in the Road Corridor 32
   2.6.2   Utility Relocation Practices 34
   2.6.3   Utility Accommodation Practices 47
   2.6.4   Emerging Practices...48
   2.6.5   Transportation Utility Corridor 52
2.7   CONCLUSION..53

3.0   RESEARCH METHODOLOGY..54

3.1   RESEARCH DESIGN AND APPROACH..54
3.2   SELECTION OF THE CASE STUDY AREA 54
   3.2.1   Site Selection 55
   3.2.2   Location of the Study Area 56
3.3   DATA COLLECTION TECHNIQUES AND PROCEDURES 58

3.3.1   Data 58

3.3.2   Secondary Data 64

3.4   DATA ANALYSIS PLAN 64

3.4.1   Data Processing 64

3.4.2   Data Analysis Methods 65

PART TWO: ANALYSIS OF INTERACTION OF INFRASTRUCTURE 66

4.0   INTERACTION OF INFRASTRUCTURE IN THE ROAD CORRIDORS IN DAR ES SALAAM ... 67

4.1   INFRASTRUCTURE LAYOUT AND PHYSICAL CONDITION 67

4.1.1   Layout and Condition of Road Network 69

4.1.2   Layout and Condition of Storm Water Drains 73

4.1.3   Water Supply Network 75

4.1.4   Sewer Management Network 76

4.1.5   Electric Power Supply Network 78

4.1.6   Telecommunications Network 79

4.1.7   Assessment of Relative Location of Infrastructure within Road Corridors 81

4.1.8   Operational and Faults Interactivity among Infrastructures in Dar es Salaam 82

4.1.9   Interactivities in Urban Infrastructure Operations 83

4.2   INTERDEPENDENCIES IN INFRASTRUCTURE DEVELOPMENT AND MANAGEMENT 87

4.2.1   Development Approaches 87

4.2.2   Infrastructure Management 88

4.2.3   Stakeholder Interviews Concerning Infrastructure Management 90

4.3   ANALYSIS 92

4.4   CONCLUSION 94

5.0   MOVEMENT AND SAFETY IN ROAD CORRIDORS IN DAR ES SALAAM 95

5.1   INTRODUCTION 95

5.2   LEVEL OF MOVEMENT AND ROAD CAPACITY 96

5.2.1   Conceptual Review of Movement within the Roadway 96

5.2.2   Movement Evaluation along Road Corridors in Dar es Salaam 100

5.2.3   Conceptual Model 101

5.2.4   Data Collection through Survey 102

5.2.5   Survey Results 104

5.3   AVERAGE CAPACITY PER LANE ON HIGHWAYS 105

5.4   ANALYSIS OF UI VERSUS MOVEMENT AND SAFETY IN THE ROAD CORRIDOR 107

5.4.1   UI Interactions versus Road Conditions 107

5.5   OVERALL SAFETY SITUATION 108

5.5.1   How Tanzania Compares with other Countries in Road Accident 108

5.5.2   Road Accidents in Dar es Salaam 114

5.6   MOVEMENT AND SAFETY HAZARDS 117

5.6.1   Design-related Movement and Safety Hazards 117

5.6.2   Surface Condition-related Safety Hazards in the Road Corridors 119

5.6.3   Road Edge and Adjacent Land-Related Safety Hazards 123

5.7   RELATIONSHIPS BETWEEN ROAD SAFETY AND ROAD CORRIDOR ENVIRONMENT 125

5.7.1   Road Corridor Environment 125

5.7.2   Safety of Road Corridors 126

5.7.3   Road surface and safety of movement 127

5.8   ANALYSIS OF THE FINDINGS 127

5.8.1   Respondents Profile 128

5.8.2   Safety of Movement Element Ratings 130

5.8.3   Average Safety of Movement 132

5.8.4 Factor Analysis of the Elements of the Safety of Movement 132
5.8.5 Analysis and Hypotheses Testing 132
5.9 CONCLUSION 133

PART THREE: INTEGRATING INFRASTRUCTURE SYSTEM IN THE ROAD CORRIDOR 135

6.0 INTEGRATED INFRASTRUCTURE SYSTEM 136

6.1 THE INTEGRATED HIGH PERFORMANCE INFRASTRUCTURE APPROACH 136
6.1.1 Characteristics of the Integrated High Performance Infrastructure 136
6.1.2 Integrated High Performance Infrastructure Requirements 137
6.1.3 Potential Benefits of Integrated High Performance Infrastructure 140
6.2 IMPLICATION OF IHPI IN THE ROAD CORRIDOR 141
6.2.1 Characteristics of an Integrated High Performance Road Corridor 141
6.2.2 Components of Road Corridor with High Performance Infrastructure 142
6.2.3 Performance Potential of Integrated Corridor with High Performance Infrastructure 143
6.3 CONCEPTUAL PLAN FOR INTEGRATED HIGH PERFORMANCE INFRASTRUCTURE 144
6.3.1 Complete Street Concept 144
6.3.2 Zoning Concept 145
6.3.3 Utility Accommodation 145
6.4 INTEGRATED HIGH PERFORMANCE INFRASTRUCTURE ALTERNATIVES 148
6.4.1 Hybrid IHPI Alternative 149
6.5 APPLICATION OF THE IHPI CONCEPT TO ROAD CORRIDOR IN DAR ES SALAAM 149
6.5.1 Adoption of IHPI for Improvement of Utilities Infrastructure 150
6.5.2 Improvement of Location of Roadside Infrastructures along the Study Area 154
6.5.3 Proposed Institutional Arrangements 154

7.0 ANALYSIS OF PERFORMANCE OF THE INFRASTRUCTURE IN DAR ES SALAAM 157

7.1 INTRODUCTION 157
7.2 CHARACTERISTICS OF RSCA MODEL 157
7.3 STRUCTURE OF THE MODEL 163
7.4 DEVELOPMENT OF A PROTOTYPE 166
7.5 PERFORMANCE EVALUATION 167
7.6 CONCLUSIONS 167

8.0 SIMULATED PERFORMANCE CHARACTERISTICS OF THE PROPOSED SYSTEM 168

8.1 INTRODUCTION 168
8.2 APPLICATION OF THE RCSA MODEL TO THE ROAD CORRIDORS 168
8.2.1 Validation of the Model 168
8.2.2 Data for RCSA Model Validation 169
8.2.3 Results and Implications 169
8.3 MOVEMENT AND SAFETY PROPOSED PLANS 170
8.3.1 Countermeasure Costs 170
8.3.2 Countermeasures 171
8.4 COST IMPLICATIONS OF THE INTEGRATED HIGH PERFORMANCE INFRASTRUCTURE 173
8.5 POTENTIAL SAFETY BENEFITS OF INTEGRATED HIGH PERFORMANCE INFRASTRUCTURE 174
8.5.1 Integrated High Performance Infrastructure Concept 174
8.5.2 Integrated Infrastructure Development 175
8.5.3 Integrated Operations and Management 175
8.6 ANALYSIS OF EVALUATION RESULTS 176
8.6.1 Evaluation of Existing UI Elements with Road Features 176
8.6.2 Evaluation of IHPI Elements with Road Features 177
8.6.3 Hypothesis Testing 182

8.7   CONCLUSION..184

PART FOUR: SYNTHESIS...185

9.0   SYNTHESIS..186

9.1   DISCUSSION OF THE RESEARCH FINDINGS...186

   9.1.1   Infrastructure Interdependences in the Urban Road Corridor......................186

   9.1.2   Movements and Safety Conditions in the Road Corridor.............................187

   9.1.3   Urban Infrastructure Development Related Conflicts....................................187

   9.1.4   Urban Infrastructure Operations and Management Related Conflicts............188

   9.1.5   Integrated Infrastructure for Sustainable Improvement of Movements and Safety...............189

9.2   CONCLUSIONS..189

9.3   CONTRIBUTION TO KNOWLEDGE...190

9.4   RECOMMENDATIONS...191

   9.4.1   Practical Implications...191

   9.4.2   Recommendations to Practitioners...192

   9.4.3   Recommendation to Policy Makers..192

   9.4.4   Areas for Further Studies...193

REFERENCES...194

ANNEXES..207

LIST OF FIGURES

Figure 1.1 Road Corridor Components 3
Figure 1.2 Road Cross-section 5
Figure 1.3 Factors contributing to accidents relative weight 7
Figure 2.1 Realms of design elements.. 11
Figure 2.2 Features in roadside zone 12
Figure 2.4 Mode 1 Rutting - shear deformation within the granular layers of the pavement 18
Figure 2.5 Mode 2 Rutting – shear deformation within the sub grade with the granular layer 18
Figure 2.6 Water affection on road layers.. 18
Figure 2.7 Load transferred to the pipes under pavement.. 19
Figure 2.8 Sources of water infiltration in pavement structure 20
Figure 2.10 Critical trench depth based on soils properties 23
Figure 2.11 Utility cut effects on pavement condition 23
Figure 2.12 Overstressing of pavement and natural materials adjacent to the trench 25
Figure 2.13 Settlement profile of poorly performing utility cut in asphalt pavement........ 25
Figure 2.14 Reflective cracking from trench reinstatement 26
Figure 2.15 Material sloughing off edges of trench.. 27
Figure 2.16 Water and wastewater interdependencies 31
Figure 2.17 Coordination between different authorities on public works projects in the US 36
Figure 2.18 Utility road corridor permit process flow chart 48
Figure 2.19 Examples of utility corridor structures with and without walkway 50
Figure 2.20 Typical TUC cross section 52
Figure 3.1 Map of Tanzania and its Road Network 57
Figure 3.2 Map of Dar es Salaam urban roads (main study site)................................... 62
Figure 3.3 Photos of the data collection sites.. 61
Figure 3.4 Most common road corridor features in the study area................................. 63
Figure 3.5 Video rating data processing 65
Figure 4.1 Existing urban infrastructure assets in the road corridors of the study area 68
Figure 4.2 Typical links and junctions in Dar es Salaam 70
Figure 4.3 Road Corridor Condition for Morogoro, Kilwa and Bagamoyo 79
Figure 4.4 Drainage problem along Mandela Road, Dar es Salaam 74
Figure 4.5 Effect of water on the road surface.. 75
Figure 4.6 Condition of the underground water pipes along Bagamoyo road 75
Figure 4.7 Underground water pipes along road corridors in Dar es Salaam................ 76
Figure 4.8 Sewer pipe leakages at Magomeni and Red Cross along Bagamoyo and Morogoro 77
Figure 4.9 Sewer pipe network under the study area 77
Figure 4.10- Relocation of electricity poles and cables... 79
Figure 4.11 Operations of telephone cables in the road corridors................................. 81
Figure 4.12 Interactivity of infrastructure on each other 84
Figure 4.13 Cut of road section for underground sewer at Ohio along Bagamoyo road 86
Figure 4.14 Road surface failure due to underground infrastructure repair along Bagamoyo road. 86
Figure 4.15 Various operators managing urban infrastructure 89
Figure 5.1 Influence of level of movement and safety to society.................................... 96
Figure 5.2 Classification of roadway capacity estimation methods 97
Figure 5.3 Generalised relationships among speed, density, and flow rate on uninterrupted-flow. 98
Figure 5.4 Interrupted flow 100
Figure 5.5 Conceptual model of effect of FOI on road capacity 101
Figure 5.6 Pictures of surveyed road sections... 104
Figure 5.7 Average capacity per carriageway along Morogoro road 106
Figure 5.8 Traffic flow along Morogoro road... 106

Figure 5.9 Conceptual presentation of the causal-effect between UI and movement and safety 107
Figure 5.10 Road accident victims in Tanzania 1975-2007... 109
Figure 5.11 East Africa fatality rates per vehicle... 112
Figure 5.12 East Africa fatality risks per population.. 112
Figure 5.13 Percentage change in accidents fatalities.. 113
Figure 5.14 Evolution 1990 – 2010 EU 25 road facilities... 113
Figure 5.15 Road traffic accidents in Dar es Salaam and in Tanzania 114
Figure 5.16 Road accidents indicating faulty road surface.. 115
Figure 5.17 Accidents percentage in the study sites.. 115
Figure 5.18 Distribution of road accidents in the road corridors 116
Figure 5.19 Road accidents and distance from road edge 118
Figure 5.20 Design of the urban infrastructure as related to safety hazards 119
Figure 5.21 Poor reinstated road section and accident on same section....................................... 119
Figure 5.22 Water layer on road surface due to water pipes defects 120
Figure 5.23 Vehicle in unprotected excavation in study area 120
Figure 5.24 Excavations within roadway for repair of underground sewer pipes. 121
Figure 5.25 Uncovered manhole on road surface along Bagamoyo road 121
Figure 5.26 20 mm Cover above carriageway in wheel path 122
Figure 5.27 Flooding on road section at Chang'ombe along Mandela road 122
Figure 5.28 Vendor on walkway at Ubungo along Morogoro road... 124
Figure 5.29 Roadside friction along road corridor.. 125
Figure 5.30 Typical scenes of pedestrian and vehicular interaction in the corridor 126
Figure 5.31 Relationship between road surface conditions and accident rate............................ 127
Figure 5.31 Distribution of respondents by level of education... 128
Figure 5.33 Roadside Infrastructure effects on safety of movement.. 129
Figure 5.34 The effect of road corridor features on safety 129
Figure 6.1 Zones that comprise a typical complete street... 144
Figure 6.2 Structure to house the utility facilities for 12 companies 146
Figure 6.3 Utility corridors with and without walkway respectively.. 146
Figure 6.4 Joint trenching... 153
Figure 6.5 Two proposed typologies of 2 lanes 2 way road sections 150
Figure 6.6 Integrated high performance road corridor with 2 lanes 2 way 151
Figure 6.7 Proposed typology of 4 lanes 2 way plan.. 152
Figure 6.8 Integrated high performance road corridor with 4 lanes 2 way 152
Figure 6.9 Proposed management structure 155
Figure 7.1 Major aspects and structure interactions in the road corridor 158
Figure 7.2 RCSA Model... 162
Figure 7.4 Information flow during RCSA 164
Figure 7.5 Interfaces for the RCSA Model 165
Figure 7.6 RCSA Model... 166
Figure 8.1 Results of road features in the selected routes.. 169
Figure 8.2 Application of the RCSA Model 170
Figure 8.3 Conceptual illustration of integrated UI development 175
Figure 8.4 Conceptual illustration of integrated operations and management.......................... 176
Figure 8.5 Variation of safety of movement and road surface condition 178
Figure 8.6 Potential benefits of IHPI to road surface condition... 179
Figure 8.7 Variation of safety of movement with the lane width.. 179
Figure 8.8 Variation of safety of movement with the shoulder width... 180
Figure 8.9 Potential benefits of IHPI to geometric design parameters 181
Figure 8.10 Potential benefits of IHPI to the traffic and control factors 188

LIST OF TABLES

Table 1.1 Factors Affecting a Crash 6
Table 2.1 Effect Ratios 14
Table 2.2 Factors Contributing to Water System Deterioration 15
Table 2.3 Quantifiers for buried pipe-road system effectiveness 30
Table 2.4 Recommended utility accommodation alternatives 49
Table 2.5 Advantages and Disadvantages of Utility Accommodation Alternatives 51
Table 2.6 Design Considerations for Utility Accommodation Alternatives 54
Table 3.1 Selected urban principal arterial roads 55
Table 3.2 Distribution of Respondents by Road Location (N=500) 59
Table 3.3 Authorities Involved in the In-depth Interviews 59
Table 3.4 Sites of the Surveillance Cameras 60
Table 4.1 Urban Infrastructure Assets along the Road Corridors 68
Table 4.2 Study sites characteristics 71
Table 4.3 Summary Inventory Survey 72
Table 4.4 Manholes on the road surface along Morogoro road 78
Table 4.5 Telecommunication Cables and Poles Network on the Road Sections 80
Table 4.6 Urban Infrastructure Proximity Rating for Morogoro road. 81
Table 4.7 Number of times infrastructure leading to failures 85
Table 4.8 Main observations from the stakeholders interview 91
Table 4.9 Correlation between proximity and fault interaction 92
Table 4.10 Analysis of variances (ANOVA (b)) 93
Table 4.11 Analysis of Coefficients (a) 93
Table 5.1 Summary of studied road links 103
Table 5.2 Summary of the Field Capacity Survey 103
Table 5.3 Vehicle Classification for Road Capacity Analysis 104
Table 5.4 Passenger Car Equivalent in the road corridor along Morogoro road 105
Table 5.5 Summary of Average Capacity per Lane on Each road (PCU//h) 105
Table 5.6 Analysis of variances (ANOVA (b)) 108
Table 5.7 Analysis of Coefficients (a) 108
Table 5.8 Distribution of fatalities and seriously or slightly injured persons 110
Table 5.9 Major contributing factors of road accidents 110
Table 5.10 Distribution of road deaths, motor vehicles and population 111
Table 5.11 Casualties by Accident Type along the Corridor 116
Table 5.12 Casualties by Vehicle Type 117
Table 5.13 Fatalities and Injuries by Most Harmful Events 123
Table 5.14 Distribution of respondents by age (N=500) 128
Table 5.15 Analysis of Variances of the Element of Safety Movement 131
Table 5.16 Average Safety of Movement Ratings 132
Table 5.17 KMO and Bartlett's Test 132
Table 5.18 Analysis of Variances (ANOVA(b)) 133
Table 5.19 Analysis of Coefficients (a) 133
Table 6.1 Distance from the edge of the carriageway 154
Table 6.2 Proposed distance of the aggressive Infrastructure away from the road edge 154
Table 7.1 State Definitions 159
Table 7.2 State Definitions of the Geometric Design Parameters 160
Table 7.3 State Definitions of the Traffic Factors and Road Furniture 161
Table 8.1 Proposed Countermeasures 171
Table 8.2 Annual Cost Saving 172

Table 8.3 Improvement Requirement and Impacts of Countermeasures................................. 173
Table 8.4 Elements of Integrated High Performance Infrastructure 174
Table 8.5 Potential Impact of IHP Road Surface Condition on Movement and Safety 184
Table 8.6 Analysis of Variable (ANOVA(b)) 183
Table 8.7 Analysis of Coefficients (a)... 183
Table 8.8 Summary of the Evaluation Results... 184

LIST OF ABBREVIATIONS AND ACRONYMS

AADT	Average Annual Daily Traffic
AASHTO	American Association of State Road and Transportation Officials
AC	Asbestos Cement
ANOVA	Analysis of Variances
APWA	American Public Works Association
BCR	Benefit Cost Ratio
BCR	Benefit Cost Ratio
BMP	Best Management Practices'
C	Commercial
CATV	Cable TV
CBD	Central Business District
CCTV	Closed Circuit Television
D	Divided roadway
D/DP	Day/Dry Pavement
DAWASA	Dar es salaam Water and Sewerage Authority
DAWASCO	Dar es salaam Water and Sewerage Company
DCC	Dar es Salaam City Council
DEGES	German Unity motorway planning and construction company
DOTs	Department of Transportation
DRAG	Demand for Road use Accidents and their Gravity
DRC	Democratic Republic of Congo
DSA	Dar es salaam School of Accountancy
DSM	Dar es Salaam
EPS	Electrophoresis Power Supply
EU	European Union
FCM	Federation of Canadian Municipalities
FFV	Free-Flow Speeds
FHWA	Federal Highways Administration
FOI	Faulty Operation Interaction
FRA	Federal Roads Administration
FS	Flat and Straight
GIS	Geographic Information System
GNP	Gross National Product
GPS	Global Positioning System
GRP	Grid Reference Point
H_2O	Water
HCM	Highway Capacity Manual
HDD	Horizontal Directional Drilling
HDPE	High Density Polyethylene
HTML	Hyper Text Markup Language
HVOSM	Highway Vehicle Object Simulation Model
I	Industrial
ICE	Institution of Civil Engineers

ICT	Information and Communication Technology
IHPI	Integrated High Performance Infrastructure
IHT	Institute of Highways and Transportation
IMS	Information Management System
IRI	International Roughness Index
JICA	Japan International Cooperation Agency
KMO	Kaiser-Meyer-Olkin
KSI	Killed and Serious Injuries
LID	Low Impact Development
MAAP	Micro computer Accident Analysis Package
MC	Municipal Council
MEM	Ministry of Energy and Minerals
MID	Ministry of Infrastructure Development
MNL	Multinomial Logit
MoE	Ministry of Education
MoH	Ministry of Health
MoHA	Ministry of Home Affairs
MoWI	Ministry of Water and Irrigation
MTI	Multiple-Document Interface
MW	Megawatts
NFP	Netherlands Fellowship Programme
NRC	National Research Council
NRSC	National Road Safety Council
NSSF	National Social Security Fund
O&M	Operation and Maintenance
ODA	Overseas Development Administration
OECD	Organisation for Economic Co-operation and Development
PE	Polyethylene
PMORALG	Prime Minister's Office, Regional Authority and Local Government
PS	Pipe Stiffness
PVC	Polyvinyl Chloride
R	Residential
RCSA	Road Corridor Safety Analysis
RCSF	Road Corridor Safety Feature
RDAs	Restricted Development Areas
RMMS	Road Maintenance Management Systems
ROW	Right of Way
SHAs	State Highway Authorities
SIDA	Sweden International Development Agency
SLS	Serviceability Limit Sate
SMD	Survey and Mapping Division
SPSS	Statistical Package for the Social Sciences
SUDP	Sub-Saharan Developing Planning
TANESCO	Tanzania Electric Supply Company
TANROADS	Tanzania National Road Agency

TAZARA	Tanzania Zambia Railway
TN	Transmission
TPDC	Tanzania Petroleum Development Corporation
Trans	Transportation
TRC	Tanzania Railways Corporation
TRRL	Transport and Road Research Laboratory
TTCL	Tanzania Telecommunication Company Limited
TTI	Texas Transportation Institute
TUCs	Transportation and Utility Corridors
TxDOT	Texas Department of Transportation
TZS	Tanzania Shillings
UD	Un-divided roadway
UGS	Utility in Good Standing
UI	Urban Infrastructure
UK	United Kingdom
ULS	Ultimate Limit State
UNECA	United Nations Economic Commission for Africa
UNESCO	United Nations Educational, Scientific and Cultural Organisation
USA	United States of America
UTM	Universal Traverse Mercator
VCR	Videocassette Recorders
WEF/ASCE	Water Environmental Federation and American Society of Civil Engineers
WHO	World Health Organisation

MAIN CONCEPTS

Camber (Cross Fall)	The transverse slope of a carriageway on a section of straight alignment
Carriageway	The part of a road used by vehicular traffic
Catchment Area	The area from which water runs off by gravity to a collecting point
Centre Line	The middle of the carriageway normally marked with a white dashed line on a paved road
Compaction	Compacting embankment by roller to increase density of soil composing embankment body. It improves the mechanical properties of soil
Cross-fall	The transverse gradient or fall across a formation of pavement
Cross-section	Section through the road construction at the right angle to the centre-line
Cut (Cutting)	Excavation in natural ground usually with graved slopes
Cycleway	A track used for bicycle traffic and usually separated from other parts of a road by kerb stone or a similar structure
Ditch (Drain)	A long narrow excavation designed or intended to collect and drain off surface water.
Drainage	The interception and removal of ground and surface water by artificial or natural means
Drainage Channel	A waterway or gutter to carry away surface water
Facility	"Facility" means any tangible thing located wholly or partially in, above or underneath the Public Road corridor, including, but not limited to, lines, pipes, wires, cables, conduit facilities, poles, towers, vaults, pedestals, boxes, appliances, antennas, transmitters, gates, motors, appurtenances or other equipment and systems. They can be divided into infrastructure assets and others.
Footpath	A track mainly for pedestrians and usually separated from other parts of a road by kerb stone or a similar structure
Guardrail	A safety barrier on an embankment or at a river crossing
Gutter	A shallow waterway provided at the edge of the road to carry surface water longitudinally
Infrastructure Assets located within a public road corridor	Infrastructure assets located within a public road corridor include roads and abutting sidewalks plus underground utilities such as water and sewer mains and overhead utilities such as telephone and power cables
Manhole	The width of a carriageway required to accommodate one line of traffic
Mitre Drain	Accessible chamber with a cover forming part of the drainage system and permitting inspection and maintenance of underground drainage pipes
Outfall	The point at which water discharges from a pipe or box culvert
Public Road corridor	Land, property, or interest therein, usually in a corridor, acquired for or devoted to different purposes and subject to the control of a public agency. It includes the area on, below, or above the corridor. Some of the different purposes in an urban public road corridor in a developing country are transportation, water and/or sewage conveyance, distribution of electricity and telecommunications, solid waste.
Road corridor environment	A set of conditions of a road corridor including physical, surface and structural conditions
Road corridor Safety	Freedom from the liability of exposure to harm or injury within a public road corridor

Road Furniture	Road or street furniture (for example traffic signs, traffic boards, traffic signals, lane markings, guardrails, and streetlights).
Road Reserve (Road corridor)	It is the area reserved for future development of the road and for the road utilities
Roadway	The portion of a road including shoulders for vehicular use
Shoulders.	Paved or unpaved part of the road next to the outer edge of the pavement; the shoulder provides side support for the pavement and allows vehicle to stop or pass in an emergency.
Side Drain	Drain beyond the shoulders, parallel to the centre-line, to take the run-off from the road surface
Slope	A natural or artificially constructed soil plane at an angle to the horizontal
Super-elevation	The raising of the outside level of the road of curves to reduce the effect of centrifugal forces and improve road-holding qualities

ACKNOWLEDGEMENT

I would like to thank God the Almighty in Jesus name, master of time and circumstances, because without Him all this would have been impossible.

I would like to thank my Promoter Prof. Meine Pieter Van Dijk and my supervisor Dr. Eddy Akinyemi, for their guidance, advice, effort and the time they spent on reading and correcting draft works and for reviewing the concepts of this research work. Their guidance and encouragement enabling me to attain my goal, and their capacity to combine critique with empathy and commitment towards this work, will always inspire me.

I am very grateful for my years at UNESCO-IHE. I would also like to thank the Government of Netherlands through my sponsor NFP who made this study possible. I would also extend my thanks to the Netherlands Embassy in Tanzania for their support throughout the time I was travelling to the Netherlands.

I am also indebted to the Government of Tanzania through my employer the Tanzania National Roads Agency (TANROADS) for their support and for allowing me to attend this programme.

Thanks to my spiritual parents Padre Felician Nkwera and Mama Mtendakazi Eldina Ntandu for their support and tireless prayers. I also appreciate the steadfast support of my husband Charles B. Masolwa and our children Angela, Maria-Natalia and Joseph. I am so much indebted to my parents mother Agnes and the late father Willibard Kayoza whose foresight and values paved the way for my education and who gently offered guidance and unconditional support. I am also very grateful to my sisters Adelina, Doresta, Anelse, Anagrace, and Jacinta and my brother Raphael Kayoza, and Arrie de Winter with their families for their moral support during this study.

A lot of thanks to the SWOV management and their library staff especially, C. D. van der Braak for her support during my study. Many thanks also to the Tanzania Technology Transfer Centre library staff at the University of Dar er Salaam for allowing me to use the library facilities during my entire research period. I also sincerely acknowledge the help received from the following ministries and organisations in Tanzania for their unreserved assistance and cooperation in the course of the fieldwork. These Ministries are the MID, MHA, DAWASA, DAWASCO, TTCL and TANESCO.

Last but not least I would like to thank Joland boots the fellowship officer at UNESCO-IHE, Cynthia van Leeuwen and my colleague Richard Buharma, Ladslaus Modestus, Oliver Nyakunga, Maria-Laura Sorrentino and N. Leonard for their friendship, challenging academic discussions and warm moments during my stay in Delft.

ABSTRACT

This study, which was conducted in the capital city of Tanzania - Dar es Salaam investigated the impact of urban infrastructure interaction on movement and safety in Dar es Salaam arterial roads. By studying characteristics of urban infrastructure interconnections, the study sought to establish the effects of urban infrastructure interactions on movement and safety in order to recommend strategies for ensuring sustainable improvement of movement and safety in the road corridors. There have been serious concerns about the effects of interaction of urban infrastructure on movement and safety, but this area has not been studied sufficiently so that relevant policy making organs and planners could be advice accordingly. The study employed both quantitative and qualitative research approaches. Data were collected through condition survey, questionnaires, interviews, documentary review, and observation using surveillance cameras. The results indicate that movement and safety are serious problems in the road corridors in Dar es Salaam. It was revealed that infrastructure within the road corridor exist in inevitable interdependencies, which significantly causes deterioration of each other while impairing movement and safety. Impairment is seriously escalated by mismatches in standards and lack of coordination of operators in planning, designing, installation including operational approaches. It is being suggested that in order to attain sustainable improvement of movement and safety, integrated high performance infrastructure has to be adopted; and for enhancing safety, a model has been developed for planners and engineers for evaluating safety compliance for existing and proposed infrastructure within the road corridors.

SAMENVATTING

Deze studie in Dar es Salaam in Tanzania onderzoekt de invloed van infrastructuur op mobiliteit en veiligheid in het wegenstelsel van deze hoofdstad. Wat is de invloed van de interactie tussen verschillende infrastructurele componenten op de veiligheid en doorloop van de hoofdwegen? Er is onvoldoende onderzoek om het beleid te adviseren en daarom worden in dit onderzoek kwalitatieve en kwantitatieve onderzoeksmethoden gebruikt om inzicht te verkrijgen. Een survey met behulp van een vragenlijst, maar ook een literatuur studie en observatie van het verkeer door middel van camera's werden gebruikt. Het resultaat geeft aan de mobiliteit en veiligheid serieuze problemen zijn in Dar es Salaam. Gebrek aan standaarden en coördinatie tussen de verschillende organisaties in de planning, design, installatie en operationele fasen dragen daar aan bij. Om een geïntegreerde infrastructuur ontwikkeling voor een duurzame verbetering van de veiligheid en doorloop van stedelijke hoofdwegen te bereiken is een simulatiemodel is gebruikt om ingenieurs en planners te tonen wat de voordelen van een dergelijke geïntegreerde benadering zou kunnen zijn.

PART ONE: INTRODUCTION, THEORETICAL FRAMEWORK AND METHODOLOGY

Part one of this thesis focuses on issues related to urban road corridors, the relationships between infrastructure systems and the effects of interactions of infrastructure on movement and safety. This part is divided into three chapters. Chapter 1 introduces the thesis and provides a brief background of the study. Chapter 2 reviews the current state of knowledge on the relationships among utilities, roads and other facilities in the urban road corridor as well as the movement and safety implications of the relationships. Added are some observations on the current state of practice with respect to the development and management of infrastructure facilities in urban road corridors. Finally chapter 3 presents the methodology of the study.

1

1.0   INTRODUCTION

This chapter presents the problem and its context. It is composed of the following sections: Background to the Research; Movement; Safety of Movement; Statement of the Problem; Research Objectives and Structure of Thesis.

1.1   Background to the Research

Infrastructure includes systems that provide water, remove waste water and waste. It also facilitates movement of people as well as goods and enables communication and exchange of information regardless of distance. Infrastructure is designed to satisfy specific social needs, but also shapes social change. In addition, infrastructure is a pervasive part of every aspect of urbanised life, and has an increasing impact on human including natural environment. It is also a base for the public wellbeing, that is, transportation systems, communication connections as well as water and energy supply networks. Those who value quality of life demand reliability of infrastructure systems and safety, among other things.

Infrastructure also plays an important role in economic development, both as a direct provider of services and as a catalyst for economic development in urban areas. Among various forms of infrastructure, road transportation system is the engine of economic activities in most urban centres. Consequently, it sustains livelihood of the people. In recent times, cities have seen a large increase in road traffic and transport demand coupled with fault interactions of urban infrastructure systems that have led to affecting movement and safety of people together with vehicles in the road corridors.

1.2   Movement

When mobility is a priority, movements are described in terms of speed-flow relationships, which describe their functionality in terms of main operational characteristics, namely, flow speed and capacity. From empirical studies such as those used in the Highway Capacity Manual (HCM 2000), it is known that various factors reduce the capacity of the road link and affect movement and safety. By implication if such factors are adequately addressed and managed, movement and safety could be improved so as to realise greater economic benefits.

Society cannot continue to rely on incremental changes to the systems. A dramatic transformation of the scope, scale and institutional architecture of the infrastructure systems is required due to urbanisation and continuously changing physical, socio-economic and other realities as well as changing demands for human environmental amenities in metropolitan areas (Schubeler, 1996). In the next fifty years, about a billion people will occupy cities in developing countries and require contemporary urban infrastructure services.

Such a challenge would also offer tremendous opportunity to effect fundamental improvements, and reduce increased congestion as well as economic loss in queues. In addition, the pressing need to find financial means for addressing the problem and infrastructure service requirements in the face of deteriorating environmental health, social, and fiscal conditions, calls for development of innovative strategies that would greatly improve life cycle performance of urban infrastructure. There is a need to develop novel ways so as to extend infrastructure lifetime and capacities, improve its functionality and expand its desirable attributes at a relatively low cost. Performance of road corridors, among other factors, can be measured by traffic movement and safety. Hindering traffic movement and safety are not the only problems faced by people in Dar es Salaam but they are major factors affecting their quality of life.

Figure 1.1
Road Corridor Components

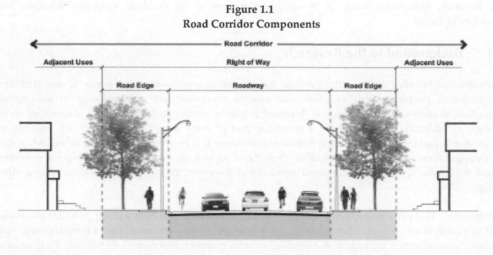

Source: Clarke, *et al.* (2008)

The term "urban road corridor" refers to both the road corridor and its interface with adjacent land uses. The road corridor includes face-to-face building separation across a road, including property outside the road corridor. Thus, each corridor functions, to varying degrees, as public space, access provider, multi-modal route and service as well as utility route. It is important to note that these four functions are not always entirely complementary to each other. Pre-eminence of various functions, often varies with each corridor type.

An urban road corridor also refers to a strip of land, which has been legally established for public purposes including direct and indirect services such as mobility of people and products, water supply and wastewater treatment systems and energy as well as communication systems. The road corridor lines separate abutting property owners from land available to the road authority for infrastructure construction and maintenance. In addition, they provide and accommodate many infrastructure systems in form of water, wastewater, storm water, roadway and utilities. In terms of movement and safety, major aspects are characteristics of the road corridors and the infrastructure systems they accommodate. The utility is a privately, publicly, or cooperatively owned line, facility, or system for producing, transmitting, or distributing communications, cable television, electricity, light, heat, gas, oil, crude products, water, steam, waste, storm water drainage, or any other similar services, including any fire or police signal system or street lighting system, which directly or indirectly serve the public [Title 23, Code of Federal Regulations, Part 645, Subpart A (23 cfr 645A)].

Focusing on the road corridor is important because the road corridor is a complex system in which roadway, utilities and other facilities are in close proximity as illustrated in Figure 1.2. It means that the infrastructure assets operate in an environment described not only by their individual inputs, outputs, and states, but also by characteristics of other infrastructure assets including certain general concerns. In such situations, the operating state and condition of each infrastructure influence the environment, and the environment in turn exerts pressure on individual infrastructure. In addition, just as little is known about effects of current concepts, methodologies, materials and actions on the magnitude, direction and sustainability of the performance of the infrastructure systems in different urban environments (World Bank, 1994).

Figure 1.2
Road Cross-sections

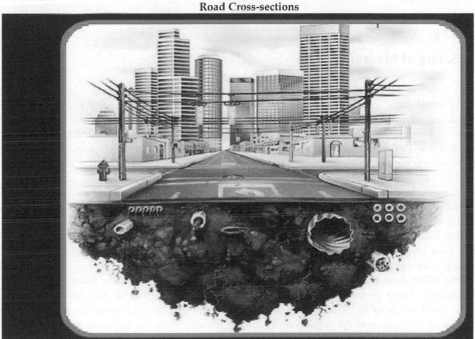

Source: Visual Dictionary (www.infovisual.info)

Little or nothing is also known about effects of characteristics and location of the infrastructure assets and how the assets interact among themselves together with the land use and other external conditions, especially in urban road corridors in developing countries. Furthermore, it is not well understood how these interactions affect movement and safety of people and vehicles.

Figure 1.3
Road Cross-section

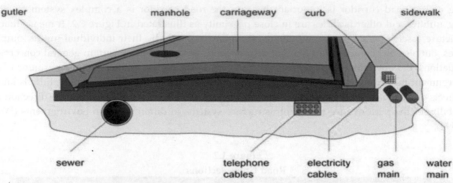

Source: Visual Dictionary (www.infovisual.info)

1.3   Safety of Movement

For every year more than 1.17 million people die in road crashes around the world. Majority of deaths, about 85 percent, occur in developing countries. Sixty-five percent of deaths involve pedestrians and 35 percent of pedestrian deaths are children such that over 10 million are crippled each year. Three quarters of all reported casualties in road accidents occur in towns and cities. It has been estimated that a million more will die and 60 million will be injured during the next ten years in developing countries unless urgent action is taken (http://www.worldbank.ord/roads/safety).

Traffic accidents in developing countries have been increasing rapidly and have, in some cases, become more deadly than diseases that affect the population (TRRL, 1991). Growth in urbanisation and number of vehicles in developing countries coupled with faulty operation interaction of the urban infrastructure have led to increased traffic accidents on road networks in urban centres. Slaughter and mutilation that occur on road corridors in the developing world everyday are events which are unacceptable in society. More people are killed and disabled in traffic accidents than in all wars that are ongoing in the world (Von Holst, 1995). Road crashes cost approximately 1-3 percent of these countries' annual gross national product (GNP). These are the resources that no country can afford to lose especially a developing economy. Road accidents are a large-scale public health problem, causing extremely high social and economic costs for society in the developing countries. Accidents happen to people from all economic groups, but more often to the poor. When injured, the poor have less chance of survival and full recovery. Accidents have substantial negative impacts on both household income and the national economy. The cost of prolonged medical care, funeral costs, coupled with loss of incomes due to disability or loss of a family bread winner can push an affected household into poverty in many low and middle-income countries. While developed countries have in general succeeded in checking and even reversing the annual number of road traffic accident fatalities, the number of fatalities in developing countries is seriously increasing and thus put a heavy burden on the already overloaded medical facilities and services. The fatalities in these countries trebled between 1980 and 2007 and will continue to increase with growing motorisation if no effective remedial actions are taken.

In an urban area, the road corridor is important for movement and safety because it is the movement environment. It is the one that interacts with human behaviour and vehicle factors before, during and after phases of a crash, affecting the probability of a crash occurring, the severity of that crash as well as the likelihood and severity of injuries occurring. Ways in which these factors may be influenced are listed in Table 1.1.

Table 1.1
Factors Affecting a Crash

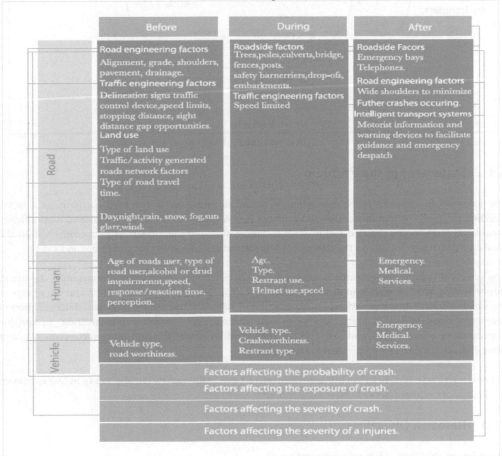

	Before	During	After
Road	Road engineering factors Alignment, grade, shoulders, pavement, drainage. Traffic engineering factors Delineation: signs traffic control device,speed limits, stopping distance, sight distance gap opportunities. Land use Type of land use Traffic/activity generated roads network factors Type of road travel time. Day,night,rain, snow, fog,sun glarr,wind.	Roadside factors Trees,poles,culverts,bridge, fences,posts. safety barnerriers,drop-ofs, embarkments. Traffic engineering factors Speed limited	Roadside Facors Emergency bays Telephones. Road engineering factors Wide shoulders to minimize Futher crashes occuring. Intelligent transport systems Motorist information and warning devices to facilitate guidance and emergency despatch
Human	Age of roads user, type of road user,alcohol or drud impairmennt,speed, response/reaction time, perception.	Age, Type. Restrant use. Helmet use,speed	Emergency. Medical. Services.
Vehicle	Vehicle type, road worthiness.	Vehicle type. Crashworthiness. Restrant type.	Emergency. Medical. Services.

Factors affecting the probability of crash.

Factors affecting the exposure of crash.

Factors affecting the severity of crash.

Factors affecting the severity of a injuries.

Source: The Haddon Matrix Adapted from Haddon *et al.* (1964)

Figure 1.4 shows the findings from road crash investigation studies in two highly motorised countries, the UK and USA. It illustrates the proportion of crashes where road environment is involved.

Figure 1.4
Factors contributing to accidents relative weight

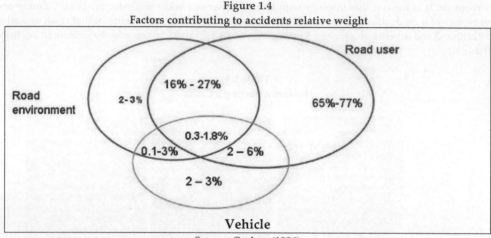

Source: Ogden, (1996)

1.4   Statement of the Problem

The focus of this research is the interaction of urban infrastructure in the road corridors as a factor affecting the road corridor movement and safety of people and vehicles in Dar es Salaam. It specifically focuses on infrastructure-related factors such as linkages among the infrastructure systems within the road corridor. It also focuses on the effects of the infrastructure interaction on movement and safety.

The principal issue in this research is related to the characteristics of the interactions among the urban infrastructure network within an urban road corridor. It is also related to the identification of effects of the urban infrastructure interactions on movement and safety in the road corridor and determination of the most cost-effective planning, designing and management strategies for ensuring sustainable improvement of movement and safety in the road corridor under the prevailing socio-economic and other characteristics that are prevalent in Dar es Salaam.

There is a serious effects of urban infrastructure interactions to the movement and safety in Dar es Salaam road corridors, which deserve to be studied seriously.

1.4.1   Research Objectives

The primary objectives of this research were to:
(i)   Identify (a) the linkages (interactions) among infrastructure assets in urban road corridors and
      (b) Relationships between the infrastructure assets to movement and safety of people and vehicles in
      the road corridor in Dar es Salaam.

(ii) Suggest alternative road corridor and infrastructure assets management strategies that can ensure
      sustainable improvement of movement and safety of people and vehicles.

(iii) Show that an integrated infrastructure development and management in a road corridor is a necessary condition for sustainable improvement of movement and safety of people and vehicles in Dar es Salaam.

1.4.2   Research Questions

Based on the research objectives the research questions addressed are:

(i)   What are the characteristics of the interactions among infrastructure assets (roadway, utilities, and other facilities) within an urban road corridor and between the assets and the corridor conditions (land-use, population and travel characteristics)?

(ii) What are the effects of the road corridor conditions on movements and safety in the road corridors?

(iii) What are the most cost-effective planning, design and management strategies for ensuring safe urban road corridor conditions under the prevailing socio-economic and other prevailing characteristics in developing countries?

1.4.3   Research Hypotheses

The research hypotheses as explained in detail in the theoretical chapter and in the chapter of the methodology of the study are:

(i)   Infrastructure assets (roadway, utilities and other facilities) in an urban corridor are geographically interdependent primarily by virtue of physical proximity as well as operational and faulty interactions among the systems.

(ii) A typical urban public road corridor in Dar es Salaam affects movement and safety, primarily due to operational and faulty interactions among the infrastructure assets in the road corridor as well as approaches to development and management of the assets.

(iii) A combination of integrated development and urban management as well as integrated road corridor infrastructure is a necessary condition for ensuring enhanced movement and safety of people and vehicles in Dar es Salaam.

1.5   Structure of the Thesis

The thesis is divided into four major parts. Part one is divided into three chapters. Chapter 1 presents the introduction. Chapter 2 reviews the current state of knowledge on the relationships among utilities, roads and other facilities in the urban road corridor and the extent of movement and safety implication of the relationships. This chapter also provides the review of the current state of practice with respect to the development and management of infrastructure facilities in urban road corridor, while chapter 3 presents the methodology of the study. Part two focuses on verifying the first two hypotheses. This part is divided into two chapters. Chapter 4 provides interaction of infrastructure and its effects on road corridor condition in Dar es Salaam as well as an approach to the development and management of infrastructure systems in the road corridor. Then, chapter 5 explains how the infrastructure interaction in the road corridor affects the characteristics of movement and safety.

Part three of the thesis deals with the third hypothesis. An integrated approach to the development and management of infrastructure systems in an urban road corridor is presented. It shows how integrated infrastructure development and management and integrated road corridor are practicable and effective. This part contains three chapters. Chapter 6 presents the characteristics of Integrated High Performance Infrastructure in the road corridor, while chapter 7 presents a model for evaluating infrastructure in the road corridor. Chapter 8 demonstrates the potential benefits of Integrated High Performance Infrastructure in the road corridor. Part four of the thesis summarises the answers to the research questions and hypotheses and draws a number of conclusions and formulates recommendations. This part contains one chapter of synthesis. In chapter 9 the results of the thesis are discussed and the last part of this chapter numbers of conclusions is drawn and recommendations are provided.

2

2.0   INTERACTION AMONG INFRASTRUCTURE COMPONENTS IN URBAN ROAD CORRIDORS: CURRENT STATE OF KNOWLEDGE

In this chapter, the current state of knowledge on the interaction among utilities, roadways and other facilities in urban road corridors is reviewed. The main issues are: Is there interaction among the infrastructure components in a road corridor? If so, what is the nature and effects of such interaction? The issues are addressed by presenting the results of studies conducted, anecdotal and empirical evidence and a methodology for quantifying structural interaction among the systems in a road corridor.

2.1   Infrastructure Components in Urban Road Corridors

The term 'urban road corridor' refers to both the road corridor and its interface with adjacent land uses in an urban area. In a built up area, the road corridor includes face-to-face building separations across a road, which include property outside the road corridor. Each urban road corridor functions, to varying degrees, as Public Space; Access Provider; Multi-Modal Route and Service and Utility Route. These four functions are not always entirely complementary to each other.

Each urban road corridor is composed of several distinct components. It is the successful design and combination of these components that create efficient and liveable communities where the end product is greater than the sum of the parts. The road corridor components can be organised into two major groups: Adjacent land and road corridors, which contain the road edge, driveways and sidewalks.

(i)   Adjacent Land

Land uses adjacent to urban roads contribute to, and greatly influence, the road character as well as function. For example, an urban road in a downtown commercial setting will have different characteristics compared to that in a suburban business park setting. Densities, orientation, quality of buildings, and quality of on-site landscaping determine characters and functions of the road corridor.

(ii)   Road Corridors

An urban road corridor includes the space, on, above and below the surface and is used for many purposes. Most importantly, it provides and accommodates many infrastructure systems in the form of water, wastewater, and storm water systems, roads and sidewalks, utilities such as gas, electricity and telecommunications along with television and computer technologies. Figure 2.1 shows the cross-section design elements of a typical urban road corridor.

Figure2.1
Realms of design elements

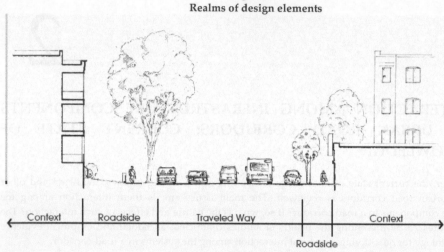

Source: Clarke *et al.* (2008)

The term roadway is often used to refer to the area of the road corridor used for vehicular travel, including cars, trucks, bicycles and transit (that is the travelled way plus the roadside in Figure 2.1). The roadway may also include a number of additional uses such as on-street parking, curbed structures such as medians and crossing islands, and utility access points. The roadway is normally centred in the road corridor, but may be offset due to topography or limited road corridor width.

2.2   Roadside Zone

The roadside zone includes an area between curbs and front property line of adjoining parcels. It should contain four sub-zones, including curb zone (edge zone), furnishing zone, throughway zone and frontage zone (as shown in Figure 2.2). These zones provide flexibilities along the length of a street for necessary landscaping, street furnishings, pedestrian through movements and roadside activities. Specific functions for each sub-zone within the roadside zone are edge, furnishing throughway, frontage and travel way zones.

2.2.1   Edge Zone

The edge zone provides an interface between parked vehicles and street furniture. The zone is generally kept clear of any objects. Normally, the edge zone has a minimum width of 0.45 m and a minimum of 1.2 m at transit points.

2.2.2   Furnishings Zone

The furnishings zone is the key buffer component between the active pedestrian walking area and the vehicle travelled way area. Street trees, planting strips, street furniture, bollards, signal poles, signals, electrical, telephone and traffic signal cabinets, signs, fire hydrants and bicycle racks should be consolidated in this zone to keep them away from becoming obstacles to pedestrians. The furnishings zone should have a minimum width of 2.1 m.

Figure 2.2
Features in roadside zone

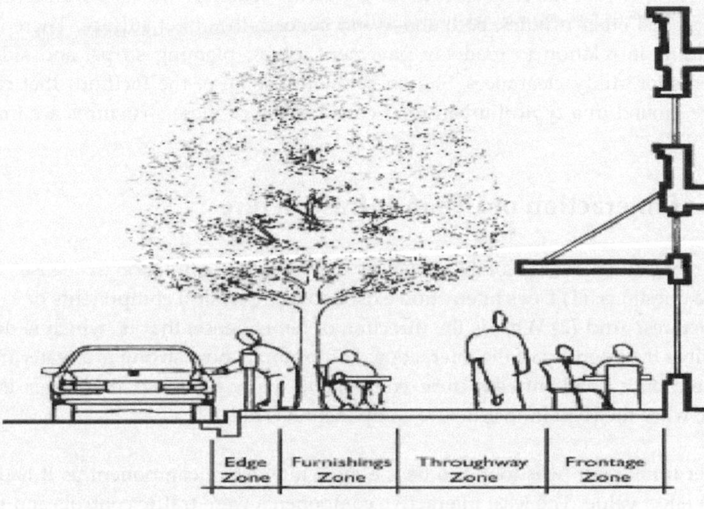

Source: Clarke *et al.* (2008)

2.2.3   Throughway Zone

The throughway zone is intended for pedestrian travel only and should be entirely clear of obstacles and provides a smooth walking surface. The throughway zone should have a minimum of six feet wide, which is the minimum comfortable passing width for two wheel chairs on a sidewalk.

2.2.4   Frontage Zone

The frontage zone is the area adjacent to the property line that may be defined by a building facade, landscaping area, fence, or screened parking area. A minimum width of 0.45 m should be provided for the frontage zone. The width of the frontage zone may be increased to accommodate a variety of activities associated with adjacent uses such as outdoor seating or merchant displays.

2.2.5   Travel Way

Travel way is a street pavement area between curbs. It includes the following key components:

i)   **Vehicle Travel Lanes**
     Vehicle travel lanes usually range from 3 to 3.6 m in width.
ii) **Bicycle Lanes**
     If provided, bicycle lanes usually have a width of 1.5 m (2.1 m including the gutter pan).

iii) **Medians**
   Medians are used as additional locations for landscaping and also serve as pedestrian refuge islands within the travelled way when needed.

One of the important functions of the road corridor is to provide space for water, sewer, electricity, street lighting, traffic signals and other utilities, both above and beneath the street surface. There are standard locations for each utility in relation to roadway pavement, curbs, planting strips, and sidewalks, and there are requirements for utility clearances. Shown below are some of the facilities that can be found above and below the ground in a typical urban road corridor. All of these structures are important and need to be managed.

2.3   Analysis of Interaction of Urban Infrastructure

Direct and indirect interaction between components was considered as a function of "cause" and "effect". Cause deals with the questions: (1) Does interaction exist between physical components or services of the considered infrastructures? And (2) What is the direction of dependence, that is, which is dependent on what? 'Effect' describes the intensity of the interaction. For example, how strong is a water infrastructure component dependent on a road infrastructure component? How well can the water infrastructure component function, when the road infrastructure component has failed?

In the analysis, water trunk main was found to be the most interactive component as it had the highest combined cause and effect value. The least interactive components were traffic controls and signs as they had the lowest cause and effect combination. The most dominant components of the system were the pavement and shoulders. These components had the highest net cause-effect value which implies that they affect the system far more than the system affects them.

- **Infrastructure is frequently the cause of failure to other infrastructure**
Water mains, roads, and gas lines (in that order) are very often the cause of damage to other infrastructures, accounting for approximately two thirds of the failures to other infrastructures in the database. See column 2 in Table 2.1.

- **Infrastructures is frequently affected by other infrastructure failures**
Gas lines, roads, electric power and fibre optic cables (in that order) are mostly affected by other infrastructure failures and are about eighty percent of the affected infrastructure.

- **Failure of one infrastructure cause failure of the other infrastructure**
Column 4 in Table 2.1 shows top six types of infrastructure involved in most of the failures. Ratios reflect the extent to which a particular type of infrastructure initiated or caused a failure of another type of infrastructure versus being affected by the failure of another type of infrastructure.

It is shown in the Table 2.1 that water mains are more frequent initiators of other infrastructure failures than the reverse. For roads and sewage facilities the direction is about equal. In addition, it shows that the most likely combinations, in decreasing order of the number of events were: gas lines and roads (16), water and gas lines (12), electric and water lines (10), and electric and gas lines (7). This may simply be a function of how frequently these facilities are co-located, or alternatively, may reflect unintended interactions that occur when these facilities are subject to external stress. This illustrative database shows that water main breaks are more frequent initiators of failure than the consequence of another infrastructure failure.

Table 2.1
Effect Ratios

Type of Infrastructure	Infrastructure (column 1) of which caused failure to other Infrastructure	Infrastructure (column 1) of which was Affected by other Infrastructure failures	Ratio of Causing vs. Affected by Failure
(1)	(2)	(3)	$(iv) = \dfrac{(2)}{(3)}$
Water mains	34	10	3.4
Roads	25	18	1.4
Gas lines	19	36	0.5
Electric lines	12	14	0.9
Cyber/ Fibre optic/ Telephone	8	15	0.5
Sewers/ sewage Treatment	8	6	1.3

Source: Ogal (2008).

2.4 Empirical and Anecdotal Evidence of Interactions on Road Corridors

In this section, we focus on the pipe-road system and present evidence of current knowledge on interactions within the utility system existing in the road corridor. Such evidence will cover the following areas

- The relationship among structural performances of buried pipes and road pavement;
- The factors relating to replacement, repair and maintenance in the serviceability of buried pipes and road pavement; and
- The collateral damage of buried pipes and pavement in a road corridor.

2.4.1 Structural Performance Relationship between Pipes and Road Pavement

(i)   Road Pavement Factors in Structural Performance of Buried Pipes
According to Serpente (1994), buried pipe deterioration is generally considered to comprise three stages. The first is the trigger event that initiates the second stage, which may be described as a decay process, which thereafter leads to collapse.

Stage1: **An initial defect**. Collapse of a sewer normally originates where an initial, often minor, defect allows further deterioration to occur.

Stage 2: **Deterioration**. Deterioration often involves the loss of support from the surrounding soil. This is discussed in more detail below.

Stage 3: **Collapse.** Collapse is often triggered by some random event that may not be related to the cause of the deterioration such as loss of support of the surrounding ground as pointed out by Serpente (1994). Therefore, it is not possible to predict when a sewer will collapse. However, it is feasible to judge whether a sewer has deteriorated sufficiently for collapse to be likely. This type of collapse has also been documented by Jones (1985), Hoffman and Lerner (1992), and WEF/ASCE (1994).

Table 2.2
Factors Contributing to Water System Deterioration

Factor		Explanation
Environmental	Pipe bedding	Improper bedding may result in premature pipe failure
	Trench backfill	Some backfill materials are corrosive or frost susceptible
	Soil type	Some soils are corrosive; some soils experience significant volume changes in response to moisture changes, resulting in changes to pipe loading. Presence of hydrocarbons and solvents in soil may result in some pipe deterioration.
	Groundwater	Some ground is aggressive or frost susceptible
	Climate	Climate influences frost penetration and soil moisture. Permafrost must be considered in the north.
	Pipe location	Migration of road salt into soil can increase the rate of corrosion.
	Disturbances	Underground disturbances in the immediate vicinity of an existing pipe can lead to actual currents or changes in the support and loading structure on the pipe.
	Stray electrical currents	Stray currents cause electrolytic corrosiveness.
	Seismic activity	Seismic activity can increase stresses on pipe and cause pressure surges.
Operation and Management	Internal transient pressure	Change to internal water pressure will change stresses acting on the pipe.
	Leakage	Leakage erodes pipe bedding and increases soil moisture in pipe zone
	Water quality	Some water is aggressive, promoting corrosion
	Flow velocity	Rate of internal corrosion is greater in unlined dead-ended mains.
	Backflow potential	Cross connection with system do not contain potable water which can contaminate water distribution system
	Operation and Management practices	Poor practices can compromise structural integrity and water quality.

Source: Hoffman *et al.* (1992)

An examination of the damage process indicates that common damage trigger events including external factors such as third party interference, like adjacent excavation, electricity and telephone poles, adjacent service failures (for example water main bursts) and road lowering. In addition there are unforeseeable loads such as impacts near traffic calming. The Federation of Canadian Municipalities (FCM) and the National Research Council (NRC) (2003) reported that road and/or adjacent systems are among the physical, environmental and operational factors that contribute to Water System Deterioration.

Further evidence of the relationships is given by Sparrow and Everitt (1977), who describe how, for sewers beneath roads, the amount of vibration caused by passing traffic affected the rate of growth of any voids, which may surround the sewer. O'Reilly *et al.* (1989) investigated the frequency of structural defects at various locations and expressed surprise that structural defects were more frequent beneath under roads. It was suggested that the high incidence of defects may stem from disturbance either during road construction or at a later date. The frequency of defects under road surface was also found to be high.

Review of literature revealed some of the factors, apart from pipes physical characteristics, that contribute to the deterioration of pipe systems (water pipes, sewer pipes, gas pipes) as given in Table 2.2.

(ii)   Surface Loading and Surface Type
The location of a pipe system will obviously affect the magnitude of surface loading to which it is subject; for sewers beneath roads the main component of such loading is likely to be that from traffic. In addition, theoretically, as illustrated in Figure 2.3 the type of pavement affects the stress on a buried pipe due to vehicle load.

Figure 2.3
Load distribution

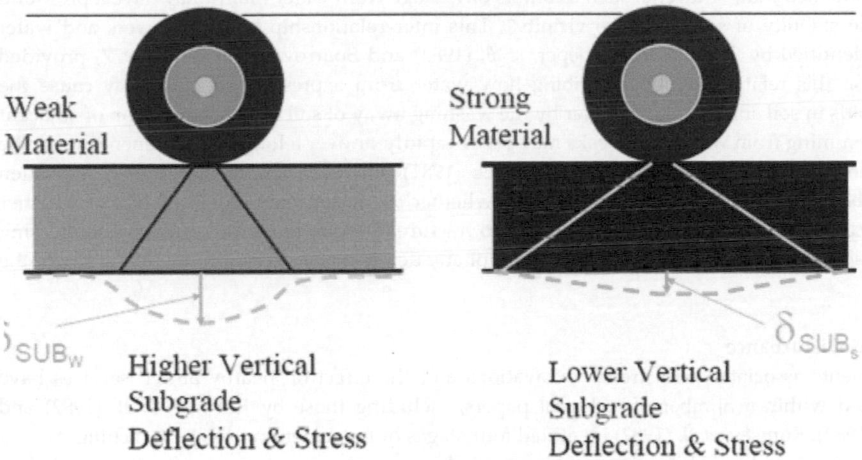

Weak
Material

Strong
Material

δ_{SUB_w} δ_{SUB_s}

Higher Vertical
Subgrade
Deflection & Stress

Lower Vertical
Subgrade
Deflection & Stress

Source: Pocock *et al.* (1980)

The Transport Research Laboratory has been responsible for a number of reports concerning the effects of traffic on road structures and buried pipes, including those by Leonard, *et al.* (1974), Pocock ,*et al.* (1980), Nath (1981) and Taylor and Lawrence (1985). Leonard, *et al.* (1974) investigated the effect of vehicle weight on dynamic loading and vibrations in road structures for eight commercial vehicles with gross weights in the range 32–44 tons. In general no obvious relationship between gross vehicle weight and dynamic loading or vibrations was found; the dynamic behaviour of individual suspension systems has the most influence on loading and vibrations. Pocock, *et al.* (1980) monitored the bending strain developed in a shallow buried pipeline due to static and rolling wheel loads. The measured bending strains were found to increase linearly with axle load, the strains for any given load tending to decrease with increasing vehicle speed. Maximum strains were always associated with pipes that had been deliberately poorly bedded. Taylor and Lawrence (1985) found that the response of an instrumented cast iron pipeline to heavy vehicles depended on several factors including the structure of the pavement, the depth of cover and the pipe bedding. Maximum strains of 13 percent of failure strain were recorded. In numerical analysis, O'Reilly, *et al.* (1989) found that increases in traffic flow appear to have been associated with a small increase in structural defect rates in principal and non-principal roads by comparison with other roads. However, the defect rate under trunk roads was found to be low, suggesting that this was probably due to stronger road pavements and greater care in design and construction at such locations. This result is partially supported by Lester and Farrar (1979) who reported a greater incidence of cracked and fractured sewers for main roads when compared to through and minor roads. The effect of loads experienced during construction has been considered by a number of authors, including Trott and Gaunt (1975), who monitored the performance of two sewers through construction to surface reinstatement and beyond. It was found that the most severe loading conditions occurred during the construction period, this being due to the heavy construction traffic traversing the pipes before the road was completed. Despite the fact that there is general recognition in the design guides that loads from construction vehicles can be significant, there is little formal guidance available for estimating such loads.

(iii)   Water Mains Bursts and Leakage
Serpente (1994) identified a 'water main break' as an example of a random event that often triggers the third stage in the process of a sewer collapse, the actual collapse itself. This view is supported by Jones (1984) who commented that, "...events such as bursts or leakage from water mains may have a profound influence on the stability of sewers in the vicinity". This inter-relationship between sewers and water mains is also identified by Sellek (1981). Koeper, *et al.* (1983) and Sparrow and Everitt (1977) provided further detail on this relationship by describing how water from a pressurised main may cause the formation of voids in soil surrounding a sewer by the washing away of soil or the compaction of adjacent soil. Failures stemming from water main leaks may occur rapidly or over a long period depending on the degree of water loss from Koeper, *et al.* (1983). Sellek (1981), however, commented that it was often difficult to establish, at the scene of a sewer collapse, whether the water main had burst first or whether the sewer had collapsed and caused the water main to fracture. Despite numerous authors commenting on this relationship, there appears to be little evidence of any detailed statistical or numerical study of the association.

(iv)   Ground Disturbance
Ground movements associated with trench excavations and the effect on nearby buried services have been investigated within a number of technical papers, including those by Rumsey, *et al.* (1982) and Symons, *et al.* (1982). Rumsey, *et al.* (1982) identified four stages of movement caused by trenching:
- As excavation proceeds and before support is installed;
- As the ground moves into contact with the support system and as the support system deflects under load;
- As the trench support is withdrawn and the trench backfilled; and
- After backfilling depending on the nature and the quality of the backfill and the ground water conditions.

Movement may continue to occur until such a time as the stresses in the backfill, and those in the surrounding ground, have equalised. Rumsey, *et al.* (1982) described procedures to minimise ground movements at each stage in the pipeline construction process. Full scale trenching trials in London clay conducted by Symons, *et al.* (1982) showed that zones of movement normal to the trench extended to a distance of 2–2.5 times the trench depth on each side, with lateral movements extending slightly further than settlements. Deflections of up to 60 mm were recorded in nearby pipelines and ground movements of up to 150 mm were measured. In similar field experiments, Rumsey, *et al.* (1982) reported horizontal ground movements of up to 120 mm and vertical movements of up to 230 mm for a 5.5 m deep, 3.5 m wide trench in a soft, sandy clay with hydraulic trench support. According to Rumsey, *et al.* (1982), the amount of ground movement at any point will depend on the distance from the trench wall, depth below ground level, geometry of the trench, soil type and properties, method of excavation, ground support, standard of construction and level of supervision. Farrar (1981) provided some measurements of differential settlement above reinstated sewer trenches on the public highway. Of the thirteen trenches studied, the surfaces of four were found to have failed within one year (two within ten days) while those remaining suffered differential settlement of up to 16 mm over one year and 23 mm over four years.

Figure 2.4
Mode 1 Rutting - shear deformation within the granular layers of the pavement

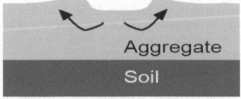

Source: Dawson (2006)

Figure 2.5
Mode 2 Rutting – shear deformation within the sub grade with the granular layer

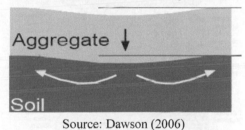

Source: Dawson (2006)

Figure 2.6
Water affection on road layers

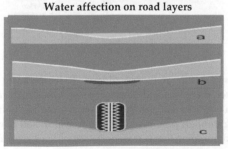

Source: Dawson (2006)

(a=water in rut affecting trafficking; b=water fed to lower pavement layers, weakening them; c=increased tyre wear)

One of the most common problems with road pavements in developing countries is rutting. Dawson (2006) shows that rutting can lead to shear deformation within the layers and sub grade as illustrated in Figure 2.4 and 2.5. In addition, ruts encourage water to soak into the pavement instead of draining off the surface and this leads to rapid pavement deterioration. Also, with rutting, water that enters the pavement may collect in a 'buried' rut in the sub grade and/or reduce the load carrying capacity of the granular layers. This reduction in capacity of pavement subjects excessive stress to the underneath structure (like pipes) to the level of being damaged due to overstressing. This is a typical phenomenon of fault interactions within the road corridor.

When this happens, more loads are transferred to the pipes under the pavement. Another common situation is water infiltrating into a pavement structure through cracks and other avenues as illustrated in Figure 2.6. This leads to weakening of the pavement layers (thus leading to transfer of higher than expected load to the buried pipes), infiltration of water into pipes as well as corrosion of the pipes. This infiltration typically occurs at manhole rings, through loose mortar and precast joints and around mainline and stub out connections as illustrated in Figure 2.7.

Figure 2.7
Load transferred to the pipes under pavement

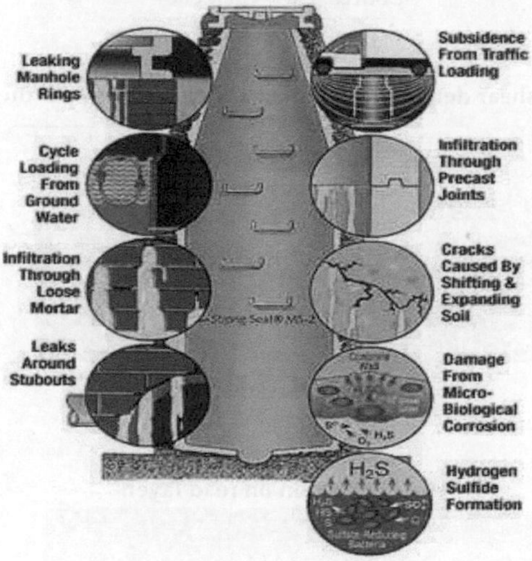

Source: Dawson (2006)

Subsidence from traffic loading, shifting and expanding soils, temperature variation and cyclic ground water loading seriously weaken manholes and other sewer system structures. Over time, ground water finds its way through fatigue cracks and weakened joints, leading to further deterioration of the structure. Also, water seeping through pavement can also transport soil particles and cause erosion and leaching of most materials and pumping that is transportation of fines around the buried pipes. Figure 2.8 shows five types of sources of water infiltration in pavement structure and are summarised hereunder:

- Seepage from the elevated surrounding soil - depends on the hydraulic gradient and soil permeability coefficient;

- Rise and fall of the underground water level - depends on the climatic circumstances and soil composition;
- The penetration of water through damaged pavement surface causes high local concentration of water in penetration areas;
- The penetration of water through shoulders; and
- Evaporation of water from the foundation soil and its condensation under the pavement structure if the pavement structure is colder than the soil.

Figure 2.8
Sources of water infiltration in pavement structure

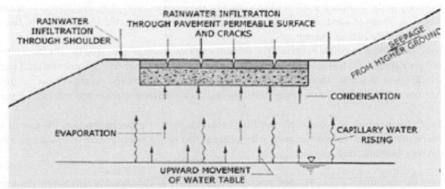

Source: Morris and Gray (1976).

2.4.2 Buried Pipe Performance Factors in Structural Performance of Pavements

One common deficiency in pipes is leakage. Leaks in water and sewer pipes buried under or near the roadway can result in the erosion of the supporting soils. Such erosion may be quickly revealed by sink holes forming in flexible asphalt pavements. Under rigid concrete pavements, large segments of the pavement may collapse at once causing considerable risk to motorists.

All structural pipeline failures are a result of inadequate soil performance. The causes of this lack of performance vary, but the ultimate failure mode is the soil. All pipelines and underground structures are soil-structure interaction systems. If the soil component of this system is poor, it will manifest itself in pavement distress, embankment settlement or collapse of the pipe.

When joints in pipelines are not watertight, there is no control over the amount of infiltration into the pipeline. With this inflow of water come fines from the surrounding backfill material. Over time, the continual loss of this material will create voids around the pipe and structures. With a flexible pipe, this may undermine the structural integrity of the pipe, resulting in a pipe collapse. With a rigid pipe, it would probably first manifest itself as localised pavement failures around the pipe or structures. These failures may result in the loss of the roadway, but as a minimum, they represent a safety hazard to the travelling public. Potholes, sinkholes and roadway collapse are examples of the type of problems, which can result from leaking storm water systems.

Whatever the reason is, once a buried pipe is broken, surrounding soil will be disturbed. This will loosen the ground and when the disturbance reaches the ground surface, local settlements will appear. Rainfall seems to accelerate the process, judging from the fact that a larger number of cave-ins occur during the rainy seasons. The difficulty is to answer more specific questions. How fast does the initial disturbance reach the surface? How can the dangerous void/cavity which will grow bigger be identified?

Unfortunately, the basic mechanism for the formation of initial voids in soil and the growth of cavity resulting in the eventual ground surface cave-in has not yet been well understood. Kuwano (2006), found that:

i)   Even a small crack or gap (5 mm wide) is sufficient to cause a cave-in.
ii)   In sandy soils, an initial small void can quickly grow, especially when the ground is saturated. Above a cavity in sand, a loosened part largely develops where the dry density decreases by approximately 10 to 20 percent.
iii)   Sand is the least resistible material. The content of a fine fraction improves the situation although not for a long time. Soil around a cavity loses confinement and when it is saturated, soil particles would yield to water. Thus, a heavy rainfall significantly affects the process of cavity growing. Above a cavity of only 5 mm in size, the loosened area vertically spreads and reaches the surface.

For sand with fines, which are commonly used for backfilling sewer pipes in practice, the rate of cavity growth seems to be slower. A loosened part, where almost half of the soil is lost, develops above a cavity, but it is not considerably large.

Najm (2005), showed that the performance of roadway pavement is significantly affected by the integrity of buried pipes underneath. He showed that damage or total loss of the pipe will result in structural damage to the pavement, excessive deflections, and roadway subsidence or collapse. Based on the results of the study, the following conclusions were made:

i)   Loss of buried pipes perpendicular to the roadway surface resulted in roadway subsidence and collapse in the vicinity of the pipe. For the typical roadway cross sections analysed in this study, the roadway subsided and then collapsed under its own weight without the application of any moving loads.
ii)   Analysis of the same typical sections for buried pipes perpendicular to traffic using stronger soils showed that the roadway did not collapse under its own weight but did collapse under moving vehicular loads. The stronger soil, under its own weight, did not collapse but suffered excessive deformation.
iii)   The analysis showed that the same failure mechanisms (roadway collapse) were observed for the pipe diameters evaluated in the study.
iv)   For the case where the pipes are parallel to traffic, the roadway surface suffered excessive deformation under its own weight. Collapse was observed in several cases under moving vehicular loads depending on the location of the moving load and the roadway profile.

2.4.3 Effects of Replacement and Maintenance of Buried Utilities and Road Pavements

All underground pipelines eventually require maintenance and repair. In an effort to maximise their water resources, public water systems increasingly require leak detection and correction. Maintenance to repair small leaks, broken valves, or leaking valve stems requires excavation to access the pipes or valves. The process of locating and exposing underground utilities for repair places other buried utilities in jeopardy of creating more leakage from the movement of unsupported, exposed joints, or directly from the excavation equipment. The trenches excavated for repair of underground utilities also collapse as shown in Figure 2.9.

A dimensionless relationship that correlates the diameter of the buried pipe (D), the depth the buried pipe (H), the critical trench depth (Z), and the side-wall cover, or minimum horizontal spacing from the trench face to the face of the pipe (X) is as follows. The ratio between the critical trench depth and the depth of the buried pipe, and the ratio between side-wall cover to pipe size is given by Equation 2.1:

$$\frac{X}{D} = 3 \times \frac{H}{Z}$$

(2.1)

Where:
X = side cover or minimum horizontal spacing, feet
D = pipe diameter, feet
H = depth of bury of the pipe, feet
Z = critical trench depth, feet

Figure 2.9
Mechanism for parallel trench collapse

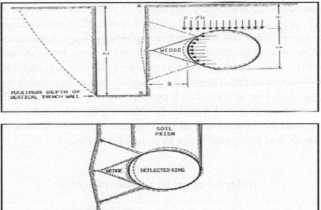

Source: Riley *et al.* (2006).

The critical trench depth (Z) is the depth at which the native soil will stand in a vertical cut without sloughing or ravelling. The engineer can estimate critical trench depth based on field experience, from a field cut, or estimated from the principles of soils mechanics. Figure 2.10 shows the critical trench depth based on soil properties. The soils mechanics method uses soil properties of unit soil weight ([γ]), soil cohesion (psf [C]), and soil friction angle of the trench wall ([φ]) as shown in equation 2.2:

$$\frac{2C}{\gamma Z} = \tan\left[45^\circ - \frac{\varphi}{2}\right] \text{ Or } Z = \frac{[2C]}{\left[\gamma \tan\left(45^\circ - \frac{\varphi}{2}\right)\right]}$$

(2.2)

Figure 2.10
Critical trench depth based on soil properties

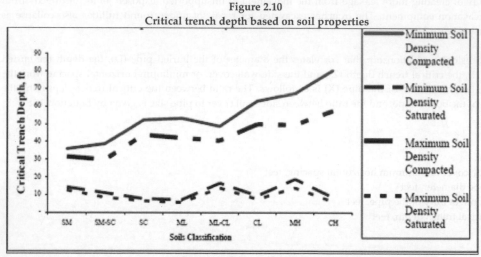

Source: Wilson *et al.* (2006)

Utility Cuts phenomenon is one issue which still remains a major concern to road authorities, and is one of the leading sources of contention between municipalities and utility service providers is the utility cut excavation and restoration. Utility cuts are made in completed pavement sections to install electric, water and wastewater utilities as well as drainage pipes under roadways. Utility cuts are also made to repair existing utilities. Utility cuts involve digging through the asphalt surface, granular base and sub-base and sub grade soil to reach buried facilities. Removal of materials used in these layers reduces lateral support to the materials in the uncut road sections. This is expected to weaken the support the excavated material provides to the pavement structure. It also causes the soil along the sides of the trench to slough into the hole reducing the support for the pavement edges. There are indications linking the digging process to loss of structural integrity and poor performance by restored utility cuts.

Figure 2.11
Utility cut effects on pavement Condition

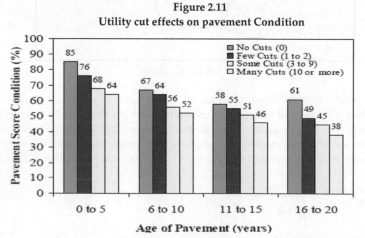

Source: Department of Public Works City and Country of San Francisco (1998)

It is the case that the pavement condition and rating decreases as the number of utility cuts made increases. For example, the pavement condition score for a newly constructed pavement is reduced from 85 to 64 as the utility cuts increased to 10 or more. In fact, several excavations in pavements by utility companies can reduce road life up to 50 percent as stated by Tiewater (1997). In some cities, millions of dollars have been spent on maintenance and repair for utility cuts made in pavements every year (APWA 1997). With the continual growth and need for repair of utilities, the effect of utility cuts on pavement performance is becoming a larger problem.

According to Schaefer *et al.* (2005), as pointed out earlier, utility cuts made in existing pavement sections to install various utilities under roadways not only disturb the original pavement, but also the base course and sub grade soils below the cut. Utility cuts in a roadway affect the performance of the existing pavement as settlement and/or heaves occur in the backfill materials of the restoration. When a utility cut is made, the native material surrounding the perimeter of the trench is subjected to loss of lateral support. This leads to loss of material under the pavement and bulging of the soil on the trench sidewalls into the excavation. Subsequent refilling of the excavation does not necessarily restore the original strength of the soils in this weakened zone. The weakened zone around a utility cut excavation is called the 'zone of influence'. Three typical pavement patch failures occur within the first year or two after the initial utility cut has been made and the pavement patch has been completed. These are summarised below.

i)   The pavement patch settles, resulting in vehicles hitting a low spot, as well as the collection of moisture, which can induce additional settlement. Typically, settlement is caused either by a combination of a poor compaction effort in natural soils or other backfill materials which have been or are exposed to wet or frozen conditions or the use of unsuitable backfill materials. A study conducted by Southern California Gas Company concluded that the top 0.6 of a metre of a backfilled excavation experiences the most settlement (APWA 1997).

ii)   The pavement patch rises forming a 'hump' over the utility cut area, particularly in winter freeze/thaw conditions due to frost action. Frost action requires three factors: (a) soils susceptible to frost (silty soils), (b) a high water table, and (c) freezing temperatures (Monahan, 1994). These factors all contribute to pavement heaving in that cold temperatures are needed for the development of the frost line, which in turn penetrates the sub grade forming ice lenses with moisture in the soil. These ice lenses continue to grow due to capillary rise and ground water table fluctuation, therefore increasing the size of ice lenses and forming visible heaves on pavements, as pointed out by Spangler and Handy (1982).

iii)   The pavement adjacent to the utility patch starts settling and failing, causing in time the patch itself to fail. This condition normally results when the natural soil adjacent to the utility trench and the overlying pavement section has been weakened by the utility excavation, as shown in Figure 2.12. This weakened zone around the utility cut excavation is called the 'zone of influence' and extends up to 1 m laterally around the trench perimeter as cited from the Department of Public Works City and Country (1998).

Figure 2.12
Overstressing of pavement and natural materials adjacent to the trench

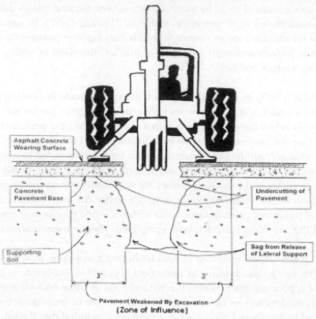

Source: Modified from The Department of Public Works City and Country (1998)

Pavement settlement occurring in and around utility cuts is a common problem that draws significant resources for maintenance. In many cases, differential settlement occurs and subsequently reduces the life of pavements in and around utility cuts as shown in Figure 2.13.

Figure 2.13
Settlement profile of poorly performing utility cut in asphalt pavement

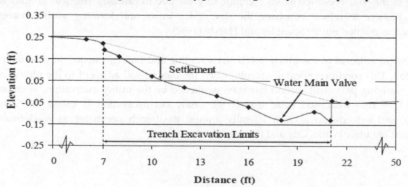

Source: Centre for Transportation Research and Education (2005).

Utility cut restoration has significant effect on pavement performance. It is often observed that the pavement within and around utility cuts fails prematurely, increasing maintenance costs. For instance, early distress in a pavement may result in the formation of cracks where water can enter the base course, in turn leading to deterioration of the pavement, (Peters, 2002). The resulting effect has direct influence on the pavement integrity, life, aesthetic value, and drivers' safety (Arudi *et al.* 2000). The magnitude of the effect depends upon the pavement patching procedures, backfill material conditions, climate, traffic, and pavements condition at the time of patching. Bodocsi *et al.* (1995) noted that new pavements should last between 15 and 20 years. However, once a cut is made, the pavement life is reduced to about 8 years. Tiewater (1997) found out that several cuts in a roadway can lower the road life by 50 percent, as shown in Figure 2.14.

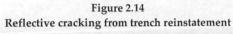

Figure 2.14
Reflective cracking from trench reinstatement

Source: Tiewater (1997).

During the excavation of a water main break, adverse conditions occur such as shown in Figure 2.15. As a result of the break, material becomes saturated and weak and begins to slough off. This in turn forms large voids underneath the existing material surrounding the cut, making adequate compaction difficult. Other problems that may arise during the reconstruction of the trench include large lift thicknesses, improper compaction, and lack of moisture control.

As an illustration of the seriousness of the utility cut problem, in the United Kingdom it is known that the direct cost of trenching and reinstatement work in highways for utilities is in excess of £1.5 billion per year, part of which is attributable to 'dry' holes (plant or equipment not found) and damage to third party assets which is estimated to be as high as £150 million. Big as they may seem, direct costs are significantly less than the societal costs such as delays to road users, disruption to businesses and environmental damage which may be as high as £5 billion per year.

There are many potential effects of trench cuts on existing pavement systems, their long-term performance and their rehabilitation requirements as pointed out by Adedayo (2007). These can be grouped in three general categories. A discussion of each category and the extent to which they were addressed by references reviewed are as follows.

Figure 2.15
Material sloughing off edges of trench

Source: Adedayo (2007).

(i)   Infiltration of Water
Water or snowmelt can infiltrate at the interface between the repaired and the existing pavement sections. There could also be infiltration through the asbestos cement of either the existing or repaired section particularly low density, high permeability asbestos cement. The water typically weakens the unbound pavement layers (aggregate base, sub-base and sub grade), and can reduce the life of the existing and repaired pavement section. Moisture-induced effects can be more pronounced in freeze-thaw environments. The impact of moisture infiltration was directly addressed in only one of the studies reviewed in Mangolds and Carapezza (1991).

(ii)   Edge Effects at the Pavement Cut
Sloughing of the sub base material and/or aggregate base adjacent to the pavement cut and inadequate compaction of the repair materials, particularly the sub grade soils, may result in the breaking down of the edge of the repaired and the adjacent existing pavement sections. This distress will allow more water to enter the pavement sections (existing and repaired), resulting in further degradation of pavement materials. The extent of this distress is dependent on the quality of the repair. Studies by Shahin *et al.* (1986) and Humphrey and Parker (1998) addressed this issue.

(iii)   Quality of Materials and Construction Procedures
If the quality of the materials and construction procedures used in the repair of a pavement cut is less than the existing pavement section, a weaker pavement section will result. With traffic loading, this can result in the initial deterioration of the pavement cut, progressing to the adjacent pavement section. Studies by Emery and Johnston (1985), Todres and Saha (1996) and Mangolds and Carapezza (1991)

characterised the placement conditions of the existing and repaired pavement sections. Theoretical studies done by Todres (1999) and Humphrey and Parker (1998) considered differences in the degree of compaction and the composition of the aggregate base or sub grade. Any of these three processes can initiate pavement deterioration, independent of the age of the existing pavement, which can then progress to either of the other processes.

In one of his researches, Adedayo (2007) went further with detailed field scale studies and numerical investigations to determine, the impact of traditional open cut-and-cover and Horizontal Directional Drilling (HDD) pipeline construction methods on flexible pavement, and Polyethylene (PE) pipe performance when installed using traditional Cut-and- Cover and HDD construction methods.

The results from the detailed studies show that the maximum change in pipe diameter of the HDD and open-cut installed pipe measured during the long-term monitoring period were approximately 1.2 and 0.75 mm, respectively. The maximum change in pipe strain measured on the buried pipe during the long-term monitoring period was approximately 4530 micro strain. Overall, the HDPE pipe performed very well over the monitoring period. The deflections and strain levels recorded were very small and well below acceptable levels, and there was no apparent deterioration of the pipe. Furthermore, it was found that there was no discontinuity created in the pavement as a result of the HDD installation. However, despite the use of best construction practices, the traditional buried pipe installation technique resulted in a noticeable hump more than 25mm in the pavement section in the vicinity of the buried pipe. The hump will result in the change of the International Roughness Index (IRI), which is a measure of the serviceability and riding comfort of the pavement. The patch pavement section started to show signs of surface distress after the first year following pipe installation. The construction joint between the repair section and the existing pavement has widened considerably allowing infiltration of water into the pavement. Infiltration, if not addressed could subsequently lead to alligator cracking and may cause structural damage. The test also showed that the repair section is not as structurally sound as the undisturbed pavement sections. Frost action was also enhanced at the location of the trench. The result revealed that the trench developed into in a significant discontinuity that will result in pavement deterioration, increased cost of maintenance and loss of life.

The numerical analysis showed that when an unsupported excavation is created within a typical flexible pavement structure, distress zones, extend laterally to a distance of about 0.54, 1.03 and 1.65 m for an excavated depth of 1.0, 1.5 and 2.0 m respectively, from the face of the trench. This equates to $\frac{x}{H}$ of 0.54, 0.67 and 0.82 for depth of excavation of 1.0, 1.5 and 2.0 m respectively, where x is the extent of the distress zone from the face of the excavation and H is the depth of excavation as pointed out by Adedayo (2007) . The vertical pressure applied on the backfill material via compaction during trench restoration does not provide sufficient lateral pressure against the face of the trench to re-compact the 'distress zone' around the utility trench. The results of the analyses suggest that a more efficient restoration technique is required to eliminate the adverse effects caused by the stress relief within the pavement structure during a utility cut excavation. It was also concluded from the numerical simulation that the compaction of backfill material has a considerable influence on the magnitude pipe deflection.

2.4.4 Excavation Damage

Excavation damage may include damage to the external coating of the pipe, or dents, scrapes, cuts, or punctures directly into the pipeline itself. Excavation damage often occurs when required One-Call notifications are not made prior to beginning the excavation, digging, or plowing activities. When the location of underground facilities is not properly determined, the excavator may inadvertently and sometimes unknowingly damage the pipeline and its protective coating.

Excavation damage can cause catastrophic failures in two ways:
(i)   It can cause immediate failure of the pipeline due to the contact between the excavation equipment and the pipeline;
(ii)   It can result in damage to pipeline coatings or dents or scrapes to the steel pipe that can lead to catastrophic failure of the pipeline at some point in the future. In this delayed failure mode, damage to coating can allow accelerated corrosion to occur which, when combined with the physical damage to the pipe steel from any accompanying denting or gouge, can result in an increased potential for future leaks, or in some cases, catastrophic failure.

The leak or failure can severely damage adjacent utilities and adjacent properties. The collapse of sewer can also cause damage to adjacent utilities. Similarly, the functional limit state is the state at which the operation and management requirements of a component are no longer fulfilled as a result of interaction between components. This implies that a change in service characteristics of a given component affects the service characteristics of the other components with which it interacts and vice-versa.

2.4.5   Functional Interactions

(i)   Quantification of the Functional Interactions
Functional condition or serviceability of infrastructure, is defined by the user expectation and judgement of the network or facility for the chosen period. Thus, to quantify the functional interactions within the system, a stepwise procedure for definition of aspects that determine the quality of services from the perspective of the users was established as follows:
• Definition of the integrated needs and desires of the users;
• Definition of the integrated problems likely to result from the interaction of the needs and desires of the users; and
• Definition of the integrated objectives, criteria and standards of addressing the problems.

The integrated needs and desires of the users of the buried pipe-road system entail quality of life-related service in terms of water, mobility and environmental sanitation. The integrated problems that may undermine attaining these needs and desires of users are:
• Inadequate access to water;
• Inadequate sanitation;
• Inadequate level of performance of roads;
• Environmental pollution; and
• Bad drainage.

The consequences and implications of the integrated problems may take three major dimensions, namely: (a) effectiveness (b) reliability (c) cost.

(a) **Effectiveness:** In this context, effectiveness measures the dexterity with which the system forestalls the manifestations of structural failure. That is, it alludes to the organisational competence in asset management of the system such that failures are premeditated and prevented before occurring. Thus, effectiveness can be described by:
• Capacity and delivery of services;

- Quality of services delivered;
- The system's compliance with regulatory standards; and
- The system's broad impact on the community.

Table 2.3
Quantifiers for buried pipe-road system effectiveness

Effectiveness	Capacity and delivery of services indicators	Quality of services delivery indicators	System's compliance with regulatory standards indicators	System's broad impact on the community indicators
Inadequate access to water	- Breakages frequency	- Hydraulic capacity	- Failure impact	- Health effects
Inadequate sanitation	- Surcharge frequency	- Hydraulic capacity	- Failure impact	- Health effects
Inadequate level of performance of roads	- Traffic congestion	- Roughness / rideability	- Failure impact	- Safety
Environmental pollution, user confidence	- Contaminant loads	- Water quality	- Failure impact	- Health effects
Drainage problem	- Flood frequency	- Hydraulic capacity	- Failure impact	- Safety

Source: Ogal *et al.* (2008)

Capacity and delivery of services can be measured in terms of frequency of water pipe breakages, frequency of surcharge at manholes, road traffic congestion, contamination loading in effluents and occurrences of flooding on the road corridor.

Surcharge at manholes may be indicative of shortfalls in the connecting sewers. Traffic congestion, in this context, are those resulting from maintenance and repair disruptions of components of the buried pipes-roads system. Likewise, flooding occurrences are those of a scale that affects mobility in road corridor.
The quality of service delivery can be measured in terms of hydraulic capacities of the buried pipes. When water pipes are blocked, the discharge is lowered resulting in a reduced rate of supply to users. Similarly, blockage in sewers and drains result in spillage or ponding of flow, conditions which may result in offensive odours. Other parameters for measuring quality of service delivery of the system include pavement ride ability and water quality. Table 2.3 shows quantifiers for buried pipe road system effectiveness.

Pavement ride ability is arguably the most representative measure of the overall pavement condition as it wholly considers the effects of the structural interactions in evaluating the level of comfort to road users. Water quality addresses issues relating to the confidence of users in water from a particular source. In the case of sewers and drainage, quality is the contaminant loading in effluent that may result in environmental pollution.

Compliance with regulatory standards is generally measured in terms of penalties borne by the service provider in the event of failure to meet set standards. The broad impact on the community of the system can be measured in terms of health effects and road safety. Health effects measures include the reported cases of water and sanitation related illnesses resulting from use of water from the system and the prevalent respiratory ailments related to inhaling gases from fuel oil combustion. Safety is assessed in terms of road accidents that can be attributed to the serviceability condition of pavement.

(b) **Reliability**: Reliability depicts the likelihood that the system effectiveness will be maintained over an extended period of time or the probability that the service will be available at least at specified levels throughout the period of the assessment. Thus, reliability can be measured in terms of:
- the time lapse between occurrence of the failure event and the commencement of repair works, and
- The time lapse between the scheduled start of maintenance works and the actual start.

(c) **Cost**: The cost dimension captures the economic implications of the failure events defined under effectiveness to the larger society within the period of assessment. Cost can be quantified in terms of:
- The length of time between commencement and completion of scheduled maintenance work;
- The length of time between commencement and completion of works to remedy unforeseen disruptions; and
- The estimated time loss due to disruptions or inconveniences from system failures that have been reported but have not been remedied during the evaluation period.

2.5   Implications of Nature of Interactions

Based on Rinaldi *et al.* (2001), a bi-directional relationship exists between the states of any given pair. A bi-directional relationship between two types of infrastructure through which the state of each infrastructure influences or is correlated to the state of the other is known as interdependency. Interdependency refers to the mutual functional reliance of essential services, e.gqw2 networked utility services on other networks, utilities, services, or auxiliary non-utility systems. While the underlying concept of dependence recognises the reliance of one system's operations on another, interdependence suggests that systems operate synergistically. Interdependency suggests that a disruption or outage in one operation will affect another, and vice versa. Two major ways that different infrastructure sectors can be connected or interdependent are spatially or functionally as reviewed by Zimmerman (2001). Spatial dependency refers to the proximity of one infrastructure to another as the major relationship between the two systems.

In this situation, a local environmental event affects components across multiple infrastructures due to physical proximity. Also, according to Zimmerman (2004), geographical or spatial interdependency involves the locational proximity of infrastructures such that a single event would be simultaneously disruptive to multiple operations. Functional dependency refers to a situation where one type of infrastructure is necessary for the operation of another, such as electricity being required to operate the pumps of a water treatment plant. Other typologies of interdependencies have been put forth. For example, Peerenboom *et al.* (2001) suggested physical, cyber, geographic, and logical categories. The two categories (spatial and functional) encompass most of the elements of the Peerenboom, Fisher and Whitfield typology. Spatial is equivalent to the geographic category and functional combines physical, cyber and logical. Infrastructure interdependencies are now recognised as both opportunities as well as points of vulnerability.

Figure 2.16
Water and wastewater interdependencies

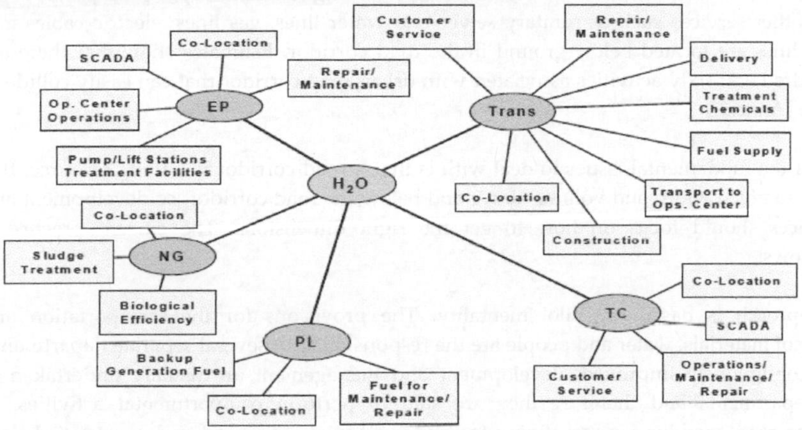

Source: Whitter (2006).

Furthermore, Zimmerman (2004) postulated institutional interdependency which occurs when the status of an infrastructure is dependent upon another with respect to policy considerations. Dudenhoetter, (2006) proposed a slightly different but similar categorisation of:

- Physical: direct linkage between infrastructures as from a supply/consumption/production relationship;
- Geospatial: co-location of infrastructure components within the same footprint;
- Policy: a binding of infrastructure components due to policy or high level decisions; and
- Informational: a binding or reliance on information flow between infrastructures.

According to Gillette *et al.* (2001), geographic interdependencies arise when infrastructure components such as water pipelines, transmission lines, gas pipelines, and telecommunication cables are co-located. Figure 2.16 illustrates some of the dependencies of the water and wastewater infrastructures with the transportation, natural gas, petroleum liquids, telecommunications, and electric power infrastructures.

2.6   Current State of Practice

2.6.1   Development and Management of Infrastructure in the Road Corridor

In the preceding section, it was shown that infrastructure facilities in the road corridor are geographically interdependent. Given this situation, one pertinent question is: Are the infrastructure facilities developed and managed as such? In other words: Are the development and management practices compatible with the requirements of such an interconnected system? This chapter therefore focuses on providing answers to these questions by examining current practices with respect to the development and management of infrastructure facilities in the urban road corridor.

In whatever way it is defined, the road corridor provides the necessary space in which to locate many above-ground functions such as roads, pavements, traffic signs and signals, landscaping and power lines. In addition, many other services such as sanitary sewer and water lines, gas lines, electric cables and telecommunication lines are located below-ground in the road corridor. Estimates show that there are more than one hundred and sixty activities associated with urban road corridor that can easily collide in both space and time (Annex I).

Furthermore, one of the fundamental issues to deal with is that a road corridor is a finite resource. It is argued that there is so much width and volume above and below the road corridor, so development and management practices should focus on how to get the right dimensions. The current practice is characterised as follows:

(i)   'Desert' approach is based on 'silo' mentality. The provisions for the transportation and conveyance of materials, water and people are the responsibility of several separate departments. Although continuously improved, development and management are usually undertaken as separate departments and therefore they are said to perform compartmental activities. In addition, the road corridor for each linear facility is planned, approved and constructed on a project or case-by-case basis. Each organisation generates routing alternatives based upon facility needs, corporate selection criteria and in light of numerous regulatory requirements. As a result, many road corridors in urban areas are basically incidental corridors. That is, although many spaces are occupied by water lines, sewer lines and other utilities, in addition to the paved road, each is planned and operated independently adjoining mainly by convenience.

(ii)   The basic premise of the second approach is that the primary purpose of a road corridor is to provide for the efficient and safe movement of people and goods with an incidental role as a utility corridor, where this is compatible with the principal role. Here the view is that the city or country that owns the road corridor has some responsibilities to see that utilities are installed in a way that minimises traffic safety issues and do not necessarily hinders travel on the road corridor. The responsibilities of public agencies generally include operating the road corridor in a way that ensures safety, traffic-carrying ability and physical integrity of their facilities. A utility's presence within the road corridor can affect these characteristics. So it is necessary that public road agencies reasonably regulate the utility's presence. In addition to laws, regulations, and ordinances, professional organisations and their publications influence utilities' rights and roles within the road corridor. For example, the American Association of State Road and Transportation Officials (AASHTO, 2007) has prepared policies and guides to distinguish good roads/utilities practices. Consequently, utilities find themselves governed by multiple agencies, many rules and laws, and various publications. Thus, the development and management of the road corridor puts primary emphasis on transportation operation and safety, maintaining the structural integrity of transport facility, preserving the aesthetic value of the facility elements, protecting the public investment in the road and associated facilities, and accommodating development or improvement of the road corridor. Thus, the use of the road corridor for the placement of utilities is often constrained and even prohibited to preserve the through traffic function.

Because of these characteristics, there are two broad areas of concern. The first of these is the relocation, replacement, or adjustment of utility facilities that fall in the path of a proposed road improvement project. This is commonly referred to as a utility relocation. The second major area of concern is the installation of utility facilities along or across the urban road corridor with the intent of occupying and jointly using this road corridor. This is commonly referred to as utility accommodation.

2.6.2   Utility Relocation Practices

An indication of the serious concern on utility relocation is the statement by Smutzer *et al.* (2004) of the Accountability, Communication, Coordination and Cooperation pg 7, who said that delays in completing road improvement projects due to utility relocation issues and conflicts create safety risks and traffic congestion, and add inconvenience and expense to taxpayers, motorists, contractors, utilities and adjacent property owners.

i)   Underlying Factors
A road improvement project typically involves four key stages: planning, designing, road corridor procurement and construction. For a successful improvement of the projects requires appropriate planning and coordination.

•   Planning
Road planning and utility planning have traditionally been carried out by two different groups' governmental agencies and investor-owned utilities. Road plans often have been developed and modified with little regard for the effects of these decisions on utility networks. Utilities have found themselves 'making do' with available and often inadequate space in street road corridors. Though utility services such as electricity, telephone, gas, and other privately owned utility services are as essential to modern life as municipally provided services water, sewer, signal, and transportation systems, their planning has not typically been integrated into the community planning process. In addition, many individual agencies and companies plan to make improvements to their facilities in isolation from others who may directly affect or be affected by these plans. In such cases, an agency may be ready to start a project when a host of other public and private agencies arrive on the scene with issues to be addressed and impacts to be mitigated before the project can proceed. In some urban areas efforts are made to ensure there is participation of everyone represented in the road corridor. Parties with facilities in or abutting the road corridor are sometimes given the opportunity to examine and consider the impact of proposals affecting that road corridor. Planning and coordination of activities related to these facilities include placement of new facilities, extension of existing service lines, replacement or upgrading the existing facilities, maintenance, and effecting service connections. Joint planning involves agencies (sometimes very diverse agencies) working together, sharing information, and developing workable and comprehensive solutions for sharing the road corridor.

•   Coordination
Coordination is a term so commonly used in road and utility projects that its meaning is sometimes diluted or confused. Coordination or active effort to share information and interact productively with others can occur in all phases of the development of a project. Benefits from coordination can be found during each phase (planning, designing, preliminary engineering, construction, operation and maintenance). Effective coordination during construction begins with better coordination prior to construction. Many road agencies engaging in project development work encounter problems in relocating existing utility facilities in the road corridor. Many of these problems are generated by poor

communication and coordination efforts. Some of the complications that result from lack of coordination or joint planning include the following:

- oInsufficient time for the utility to perform relocation design;
- oSlowness of utilities in performing their relocations;
- oShifting project and utility schedules or priorities;
- oSome project or utility plans which have omissions or errors;
- oLate changes to the existing plans;
- oAgency or utility change of policies without notification; and
- oLack of utility 3 to 5 year capital improvement plans.

These complications lead to substantial undesired costs, project construction delays, unsafe conditions, and difficult working relationships.

A comprehensive approach to infrastructure planning has to a large extent been hindered by the varying institutional structures, operating policies of public and private utilities and road agencies. With some utilities developed as government service and others as private enterprises, different approaches have evolved for planning and operating these systems. Public water supply, sewer and drainage systems were the first utilities to be installed in communities and became government-operated enterprises. Telephone, telegraph and electric supply systems followed as private or investor-owned utility services. These newer utility services were granted franchises to operate within communities. Although bound by certain restrictions, they were left to plan and develop systems for providing their services as private enterprises. As communities evolved, utility planners generally proceeded with minimal interaction with government planners.

In addition, the phase at which utilities are involved varies. For example, in some states utilities were first involved in the following phases: at the project concept level, pre-environmental, preliminary line and grade inspection, preliminary design, approval of road corridor plans, and at either 10, 25, 50, or 60 percent completion of design. The nature of the organisations also mattered. In general, State Highway Authorities (SHA's) have more specialised departments and personnel to assist in the road project development process than local departments. SHA's regularly develops road construction programmes, plan utility coordination meetings, and phase submission of plans (for example, plans may be distributed to utilities for review when they are 30 percent complete, 60 percent complete or 90 percent complete). SHA's also have specialised personnel such as utility engineers to deal with utility related matters. Local agencies have smaller organisational structures, with possibly one staff responsible for several different aspects or phases of a project. Except for urbanised areas with an established transportation planning process, local departments may not have project plans that extend beyond a couple of years. Usually the local agency does not have a utility engineer or a regular process for incorporating utilities into the project development effort. Many times, only final plans are prepared and distributed with no opportunity for utility review of preliminary design plans. Lack of local resources to deal with utilities and road corridor issues is compounded by limited road corridors typically found in urban areas. In addition, traffic congestion and intensively developed areas add constraints and urgency to any disruption of the road.

A corollary is that the approach to coordinating work with utility companies varies from place to place. In some places, a collaborative approach is used (as shown in Figure 2.17) for example, there are new, collaborative approaches to managing utilities within the road corridor emerging especially in Canada, Europe and New Zealand. Features of these approaches include a high level of commitment from senior

Figure 2.17
Coordination between different authorities on public works projects in the US

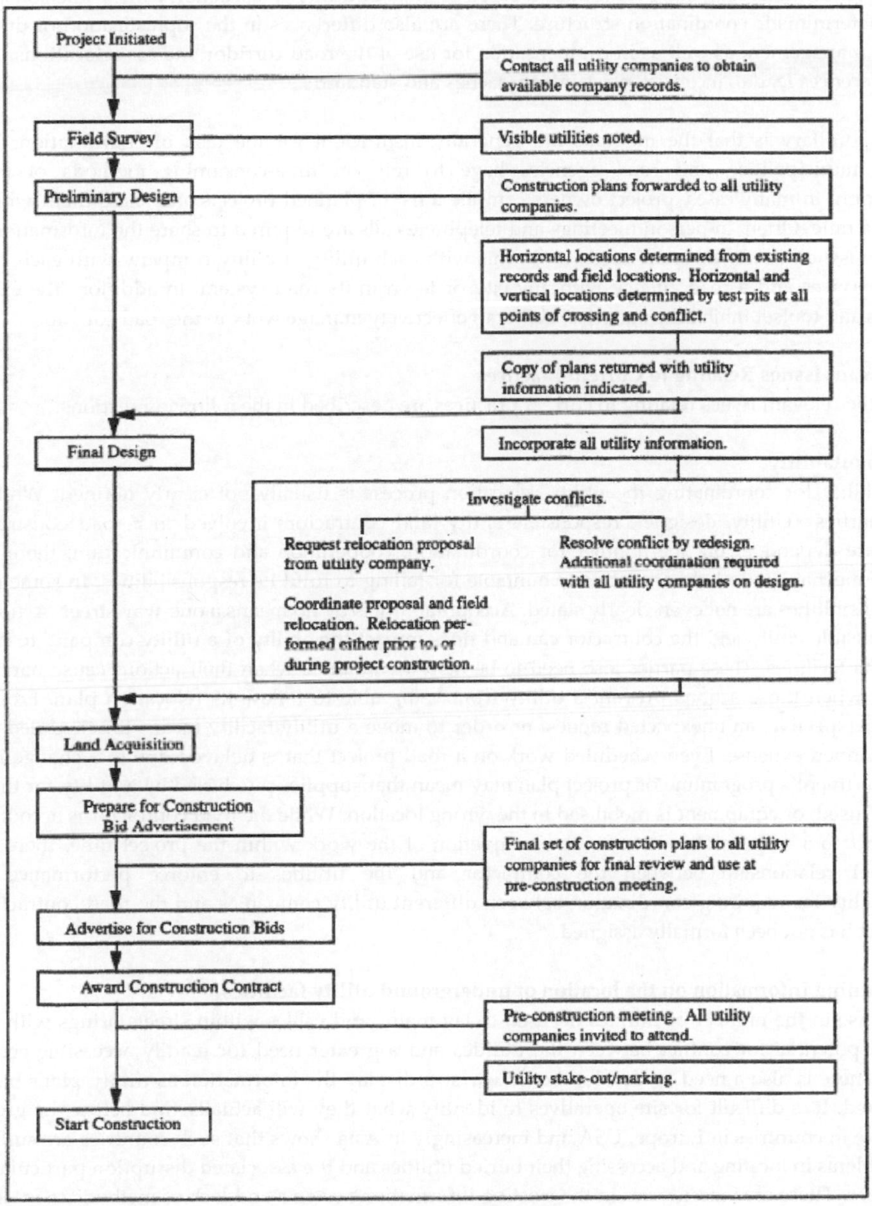

```
┌─────────────────────────────────────────────────────────────────────────┐
│  ┌──────────────────┐                                                     │
│  │ Project Initiated │                                                    │
│  └──────────────────┘        ┌──────────────────────────────────────┐    │
│         │         ───────────│ Contact all utility companies to obtain │  │
│         │                    │ available company records.              │  │
│         ▼                    └──────────────────────────────────────┘    │
│  ┌──────────────┐            ┌──────────────────────────────────────┐    │
│  │ Field Survey  │───────────│ Visible utilities noted.               │   │
│  └──────────────┘            └──────────────────────────────────────┘    │
│         │                    ┌──────────────────────────────────────┐    │
│  ┌──────────────────┐        │ Construction plans forwarded to all utility │
│  │ Preliminary Design│───────│ companies.                             │   │
│  └──────────────────┘        └──────────────────────────────────────┘    │
│         │                    ┌──────────────────────────────────────┐    │
│         │                    │ Horizontal locations determined from existing │
│         │                    │ records and field locations. Horizontal and │
│         │                    │ vertical locations determined by test pits at all │
│         │                    │ points of crossing and conflict.       │   │
│         │                    └──────────────────────────────────────┘    │
│         │                    ┌──────────────────────────────────────┐    │
│         │                    │ Copy of plans returned with utility    │   │
│         │                    │ information indicated.                  │   │
│         │                    └──────────────────────────────────────┘    │
│         ▼                    ┌──────────────────────────────────────┐    │
│  ┌──────────────┐            │ Incorporate all utility information.   │   │
│  │ Final Design  │───────────└──────────────────────────────────────┘    │
│  └──────────────┘                                                         │
│         │        Investigate conflicts.                                   │
│         │   ┌──────────────────────┐  ┌────────────────────────────┐     │
│         │   │ Request relocation proposal │ Resolve conflict by redesign. │ │
│         │   │ from utility company.     │ Additional coordination required │
│         │   │                           │ with all utility companies on design. │
│         │   │ Coordinate proposal and field │                          │   │
│         │   │ relocation. Relocation per- │                            │   │
│         │   │ formed either prior to, or  │                            │   │
│         │   │ during project construction. │                           │   │
│         │   └──────────────────────┘  └────────────────────────────┘     │
│         ▼                                                                  │
│  ┌──────────────────┐                                                     │
│  │ Land Acquisition  │                                                    │
│  └──────────────────┘                                                     │
│         │                                                                 │
│  ┌──────────────────────────┐                                            │
│  │ Prepare for Construction  │                                           │
│  │ Bid Advertisement         │                                           │
│  └──────────────────────────┘  ┌──────────────────────────────────────┐  │
│         │                      │ Final set of construction plans to all utility │
│         │                      │ companies for final review and use at  │   │
│         │                      │ pre-construction meeting.              │   │
│         ▼                      └──────────────────────────────────────┘  │
│  ┌────────────────────────────┐                                          │
│  │ Advertise for Construction Bids │                                     │
│  └────────────────────────────┘                                          │
│         │                                                                 │
│  ┌────────────────────────────┐                                          │
│  │ Award Construction Contract │                                         │
│  └────────────────────────────┘ ┌─────────────────────────────────────┐  │
│         │                       │ Pre-construction meeting. All utility │   │
│         │                       │ companies invited to attend.          │   │
│         │                       └─────────────────────────────────────┘  │
│         │                       ┌─────────────────────────────────────┐  │
│         │                       │ Utility stake-out/marking.            │   │
│         │                       └─────────────────────────────────────┘  │
│         ▼                                                                 │
│  ┌──────────────────┐                                                    │
│  │ Start Construction │                                                   │
│  └──────────────────┘                                                     │
└─────────────────────────────────────────────────────────────────────────┘
```

Source: Fairfax Country (2003).

management and cooperation in the development of policies, standards and procedures. In Canada, the Common Ground Alliance is growing and focusing its efforts on the avoidance of damage to underground utilities. Most road authorities are developing master agreements with utility companies to manage common concerns more efficiently. In some places, authorities have a utility coordinator or district level utility coordinating committee focused on road or utility construction work while some have a more deterministic coordination structure. There are also differences in the sophistication of drawing and data management as well as in compensation for use of the road corridor and for relocations. These wide differences lead to inconsistencies in approaches and standards.

Another corollary is that the methods are generally insufficient for the task of coordination. Many utilities, municipalities, and road agencies have to rely on time-consuming methods of project coordination. In many cases, project owners compile a list of planned projects and share them with one entity at a time. Often, in-person meetings and telephone calls are required to share the information and to resolve issues. A city or town needs to do this with each utility, a utility company with each city or town it services and a road agency with any city or town in its road system. In addition, the existing processes and toolset inhibit the agencies' ability to effectively manage work in the road corridor

ii) Relevant Issues Relating to Current Practices
Some of the relevant issues relating to current practices are described in the following sections.

- **Accountability**

Responsibility for coordinating the utility relocation process is usually not clearly defined. While the several parties. (Utility, designer, responsible entity, and contractors involved in a road construction project) are dependent on one another for coordination, cooperation and communication, there is no adequate mechanism to hold an entity accountable for failing to fulfil its responsibilities. In some cases, the responsibilities are not even clearly stated. Accountability is by no means a one-way street. Actions by the responsible entity and the contractor can and does impact the ability of a utility company to timely relocate its facilities. These parties also need to be held accountable when their actions cause harm to a utility or when those actions prevent a utility from being able to follow its relocation plan. From the utility's perspective, an unexpected request or order to move a utility facility means unscheduled work and unplanned expense. Even scheduled work on a road project that is delayed due to a change in the road department's programme or project plan may mean that supplies purchased by a utility for that job cannot be used, or equipment is mobilised to the wrong location. While the road contractor is accountable by contract to a responsible entity for the completion of the work within the project time, there is no contractual relationship between the contractor and the utilities to enforce performance. The responsibility for requiring coordination between different utility companies and the road contractor on the project has not been formally assigned.

- **Obtaining information on the location of underground utility facilities**

The increase in the number of utilities licensed to lay mains and cables within streets brings with it the increased potential for conflict between the utilities and a greater need for readily accessible accurate records. There is also a need to develop better ways to display the information as utility plans become complicated. It is difficult for site operatives to identify what they will actually find below the ground. Experience in countries in Europe, USA and increasingly in Asia shows that such countries are suffering from problems in locating and accessing their buried utilities and the associated disruption particularly in urban areas. There are inaccuracies in the existing information systems and lack of methods to integrate, share, reuse and effectively communicate knowledge by owners of underground assets that require more excavations to locate the underground assets, to avoid unnecessary traffic congestion and increased costs for operation. Unnecessary damage to underground assets results in increased costs, injury and or possible death of workers and loss of service to consumers, both business and domestic.

- **Nature of the coordination among public agencies, designers and utilities**

Road improvement projects move through several development phases, i.e. engineering report that determines the scope of needed improvements, data collection survey, engineering design, determination of the land needed for the improvement, formal acquisition of that land and finally construction contract preparation. As the projects move from phase to phase, project management at both agency and utility companies is often passed from one person or department to another. In those instances, no one person or department retains responsibility or control throughout the entire process. It is the operational characteristics of agencies, utility companies and design consultants that create this situation. Project continuity and responsibility are compromised as contact person's change within these organisations. Many urban road projects take multiple years to proceed through various phases. Installation of new utility facilities may occur in the road corridor during these phases. Information on these installations, which require permits, is not always known to the designer.

- **Number of relocations of utility**

Utility companies have been maintaining facilities within and adjacent to road corridors since the late 1800s. Beginning with distribution of basic municipal facilities (water, sewer and power), these road corridors now contain natural gas, communications and cable television facilities. Since more than 90 percent of the roads currently in use were built long time ago, many of these roads have insufficient road corridor to handle growing traffic demands and the proliferation of utility services. The underground environment has become increasingly congested as more and more utility facilities compete for limited space within and adjacent to the road corridor. As demand for the finite space in existing road corridor increases, the difficulty and cost of adding new utility facilities and relocating the existing ones also increase. At the same time, utility service interruptions may add to public discontent with overall road construction. It is, therefore, essential for planners, designers and builders of road projects to minimise or avoid utility facility relocations.

Road designers have little motivation to avoid utility facility relocations under the typical road design process. Designers are often faced with very tight schedules for completing road designs, which leave little time to explore alternatives that could minimise the need to move utility facilities. Efforts to "design around" existing utility facilities to avoid relocation often involve consideration of several alternatives. This extra work extends the design time and increases the design budget. When the project design is based only on where the utility facilities might be, or where they ought to be, the likelihood of encountering an undocumented facility during construction is much higher. When the impact on utility facilities is not considered early in the design process, delays are likely, either due to redesign of the road work to address relocation of utility facilities or during construction while waiting for utility companies to finish their design, land acquisition and relocation work. The costs of relocating utility facilities increase significantly if not considered during the early design process. This is especially so if they are discovered after construction has begun. The utility company must have time to prepare construction drawings, to obtain the required materials for relocation and to mobilise its crews for traffic control and construction.

- **Utility relocation coordination process during construction**

The construction phase of a road improvement project begins after all the design, road corridor acquisition and the letting process have been concluded with the award of a contract to the successful low-bid contractor. At that time, all of the previously unrelated parties of a project are first put together as a team to build the project. In this sense, the project team includes the designer (agency or consultant), agency's project engineer and staff, the contractor (along with material suppliers, subcontractors and craft workers) and the utility companies (represented by both their engineering and field construction staff). During the bidding process, the contractor must rely on information about the status and timing of the utility facility adjustments that may be based on minimal detail and assumptions since utility company work plans and proposed construction schedules are not usually a part of the bidding documents. While the contractor is contractually responsible to the agency for the successful completion of the project, the contractor has no contractual relationship with the utility companies that are involved with the project. Coordination among the members of the project construction team is required for the successful completion of the project. The road contractor's schedule and costs may be affected by utility company relocation schedules that are different from the anticipated. On the other hand, a utility company's relocation activity might be impacted when the contractor's schedule changes. Improved communication is needed among all parties. Better coordination processes are particularly needed for major projects that have significant relocation of utility facilities. Some agencies tend to use a formalised communication process known as "partnering" on many contracts in recent years to maintain communication among all parties to the construction project. The initial meeting and continuing project meeting keep all parties informed as the project evolves. Partnering can aid communication among the utility companies involved with the project that may have work plans that need to be coordinated. This process puts all the individual entities of the construction project together as a team to manage the project in a manner that benefits all parties.

- **Sufficiency of road corridor to accommodate utility relocation work**
Road construction cannot often proceed until utility companies move their facilities out of the way. Those facilities will automatically be moved to different places. This is true whether an agency acquires enough road corridors to accommodate both the road and utility facilities, or whether each party acquires its own road corridor and easements. Until there is sufficient space for both parties, construction of the road or improvements cannot proceed. Utility companies have the right to place facilities on road corridor at no cost. However, they generally must relocate at their own expenses if needed during construction, maintenance or safe operation of that road by the agency. This relationship between the agency and utility companies has existed for nearly 100 years. It began when utility companies gave up certain franchise rights in exchange for the right to occupy road corridors free of charge.

Because road corridor decisions have significant impact on the relocation of utility facilities, it may now be the right time to examine the agency's approach to road corridor acquisition. Should the road corridor be kept to the minimum width and expense needed to accommodate the road and shared with utilities only if space is available, or should the road corridor be designed to accommodate both road and utility company needs? Some utility companies prefer to place their facilities on their own road corridors or easements rather than use road corridors. This is generally a complex issue without clear answers. Each road project may require a different solution. If the agency buys enough road corridors to accommodate both the road and utility facilities, the agency would have higher initial costs to build the road. This would reduce funding available for road improvements. Locating within the road corridor, utility companies, however, could face future relocation costs if and when it becomes necessary to move the facilities to accommodate new road needs. A related issue is the agency's policy that generally prohibits the placement of utility facilities in the road corridor for limited access road. A limited access road is either a freeway with limited points for entry and exit or a road that has limited intersecting roads and connecting driveways. The responsible entity's policy stems from a time when most limited access roads were rural in nature, with high speeds and few intersections or driveways. Utility facilities were excluded from these roads as a safety measure since drivers do not expect slow-moving or parked vehicles as might occur for the maintenance of utility facilities. However, the designation of limited access road in heavily developed urban areas is not so much a safety issue as it is a tool to control proliferation of driveways and resulting turn movements that add to traffic congestion. In some cases, utility facilities are already located in a road corridor that the agency might wants to convert to a limited access road. In these situations, the agency normally wants all utility facilities relocated off the road corridor. Delays to the improvement of these roads may occur because a utility company either requests accommodation within the road corridor of the limited access road or proceeds to acquire its own road corridor on private property.

- **The road corridor acquisition process**

Acquisition of a new road corridor is a process that takes time, even when property owners have agreed on the project and proposed road corridor limits. If condemnation is needed, it becomes much more time consuming and adversely affects both the agency and utility companies. The responsible entity generally buys only enough road corridors to accommodate the road work. If a utility company cannot be accommodated within that space, delays may occur because the utility company must acquire its own easement or road corridor on private property, separate from the road corridor. Utility companies cannot proceed with their own land acquisition until the proposed new road corridor limits are firmly established. If the road corridor limits are still subject to change due to design reasons, concessions to property owners or condemnation proceedings, then the utility company is unable to proceed to any significant degree with its own design and property acquisition. In fact, property owners will often not negotiate with utility companies until all road corridor issues have been settled by the agency. If the agency had to resort to condemnation in acquiring a particular land parcel, utility companies would often face the same situation in dealing with that property owner. In that instance, two lengthy condemnation proceedings must be completed before the road can be built. At the time when the agency has finally secured the entire needed road corridor and is ready to proceed with road construction, the utility company road corridor acquisition process may just be starting and can significantly delay road improvements. A road project requiring the acquisition of a hundred different properties can be delayed for years by a lengthy condemnation process for just a single property. Until all land is acquired, the project may not be constructible from the perspective of the road contractor or utility companies.

- **Contents of road improvement contract documents**

The final product in improved coordination among the agency, its designer and utility companies is a utility company work plan for relocation of its facilities. Each utility company coordinates its relocation needs with the road design and prepares a relocation work plans. These plans are submitted to the agency for review and coordination with other utility companies. These work plans are currently submitted to the agency in a variety of forms, from simple free-hand drawings to complete computer-drafted drawings. The agency then transforms these drawings into words and includes a brief description of each utility company's relocation plan in the road construction contract documents. Currently, an all-inclusive utility relocation drawing is not developed or included in the final road plans for bidding and construction.

Road corridor preparation includes: (a) Staking the road corridor limits; (b) clearing trees and bush; (3) demolition and removal of buildings, driveways and other objects in the way of new construction; and (d) preliminary road grading and earthwork activities. Utility companies could proceed with their relocation work sooner if the road corridor was staked, cleared and graded earlier. While the agency sometimes stakes the road corridor in advance, this is not always the case. The agency sometimes has its road contractor clear those parts of the road corridor needed for utility company work but not needed for road work. However, this again is not always the case. Currently, a utility company is responsible for all aspects of road corridor preparation if it owns road corridor has an easement on private property, or places facilities on the road corridor unrelated to a road improvement project. This would include hiring a surveyor to mark the limits of the road corridor, clearing trees and bushes and demolishing and removing buildings or other objects, along with all grading and earthwork, erosion control, environmental permits, clean up and re-vegetation activities. However, when relocation of utility facilities is needed to accommodate a road project, the situation is more complicated. As part of the road construction contract, the agency normally pays a contractor to do staking, clearing, demolition and grading needed for the road work. A potential source of delay and disruption for a road contract is eliminated if a utility facility is moved before a road contract is awarded. However, to move quickly, the utility company must do staking, clearing and grading work that would otherwise have been done by the road contractor. In many instances, such work by utility companies is not eligible for reimbursement.

To further complicate matters, staking the road corridor to accommodate the relocation of utility facilities is likely to result in additional expense of staking the road corridor a second time, to accommodate the road work. Some areas of the road corridor may also need to be cleared to accommodate the relocation of utility facilities, but they are outside the limits needed for road work. The costs of the road project are increased if the road contractor does this additional clearing. Planning and coordinating the construction of a major road project that also involves relocation of multiple utilities is a complex process. This process also presents significant risks for road contractors, who have little or no control over the relocation of utility facilities in most road contracts. Because there are presently no practical mechanisms to hold utility companies accountable for their delays, the road contractors at times cannot complete the road work until all utility companies have relocated their facilities. The process is also prone to conflict because the contractors and utility companies must work in the same area at the same time, each with their own schedules and issues. The different crews tend to get in each other's way. This could lead to delays, extra expenses and work for all parties, and additional inconvenience for the public. Even when all parties work closely together, an unexpected delay by one utility company may result in a cascading series of delays, rework and extra costs for the other utility companies and the road contractor.

- **Role of responsible entity in managing the road corridor**

Judicious management of the road corridor is the key to resolving most of the issues outlined in this report. This, in part, revolves around whether an agency should acquire extra road corridor for utility facilities when those facilities must be moved to accommodate a road project. This is a key fundamental public policy issue. Utility companies and agencies both ultimately exist to serve public interests. Utility companies have the authority to acquire easements and road corridors separate from any control or involvement by the agency. If a utility company chooses to acquire its own easements and road corridor, an agency has no control over that acquisition. Multiple utility companies acting independently to acquire their own easements or road corridors to accommodate a road project is not a cost effective or efficient process. While utility companies serve public interests, they do not necessarily all serve the same public interests and may even compete with one another in some situations. If responsible entity acquires all of the land needed for both the road and utility facilities, it should be in a better position to manage the overall process, including relocation of utility facilities.

If an agency does not adequately manage the road corridor, utility companies are likely to compete with one another for the "best" location in the road corridor. Coordination among utility companies is sometimes lacking. Some utility companies do not provide adequate information on their relocation plans or proceed with work that differs from their submitted relocation plan. There may also be a lack of information on where utility facilities are located within the road corridor. This lack of information occurs, in part, because an agency has not compiled a database about the placement of utility facilities within the road corridor even though utility companies must obtain permit to install or service facilities within the road corridor. It also occurs, in part, because some utility companies do not always follow their own plans for placing facilities or in keeping records of the actual placement locations.

- **Processes for dealing with conflicts of unexpected utility facilities**

Even with additional effort and enhanced planning during the design process, problems may still be encountered during construction due to unknown utility facilities, abandoned utility facilities or additional utility facilities that may now need to be adjusted, relocated or abandoned. These facilities represent out-of-service utility lines that have been forsaken in a place. The facilities are generally of unknown origin and are attributed to either a lack of records indicating their presence or the original owner being out of business or otherwise unavailable to locate these facilities. Abandoned facilities sometimes contain useful products, and when found, they create a potentially hazardous situation. Encountering an abandoned facility during construction can mean a major delay as the facility waits to be identified, removed, or sealed. Since abandoned and active facilities are often near each other, the abandoned lines may be marked as active and vice-versa, leaving the active facility vulnerable to potential damage.

There is no defined procedure for handling unknown or abandoned utility facilities or additional utility facilities that now need to be adjusted, relocated or abandoned. This potential problem requires a plan for compensation to be in place. This will be done by the contractor (if appropriate), who will adjust the schedule as needed and will provide the utility company some response to show that there is a need for relocation.

iii) Innovative Measures and Practices in Europe and US

Recently, a number of innovative measures and practices have emerged based on a recent report on FRA (2002). The following is a summary of current road corridor and utilities best practiced in several industrialised countries:

- **Protected road designation**

In England, the road agency has the power to designate a road as protected, precluding new utility installations. This designation applies to all motorways and some trunk roads. Where a road with utility installations is designated as protected, utilities may repair existing facilities and make service connections, but may not expand or replace facilities without moving outside the road corridor. This protection is primarily for safety and operational efficiency.

In Germany, the road agency often buys extra road corridor for utilities. Road corridor cannot be purchased solely for utilities in England, but sometimes a replacement road corridor is included in negotiations when the utility company has a property right. When State DOTs acquire road corridor in the United States, FHWA encourages them to consider consulting the affected utilities and acquiring sufficient road corridor to accommodate utility needs. In addition, AASHTO's subcommittee on road corridor and utilities considers the acquisition of road corridor for utilities purposes to be the best practice. When they intend to permit utilities to occupy road corridor, some state DOTs consider this use in determining the extent of road corridor needed for the project and acquire additional road corridor solely for utility purposes. They may keep the acquired road corridor, or sell it, lease, or otherwise convey it to utilities.

• **Damage prevention**

Excavation activity causing damage to underground utilities is a problem in Europe. Utility companies in Germany are responsible for identifying their underground facilities and providing this information to road contractors before excavation. If utility companies cannot provide this information, they must physically uncover facilities at their own expenses to obtain the information. Germany is considering adopting a nationwide uniform documentation system for all utilities. Efforts are also made in the Netherlands to avoid damaging underground utilities during excavation. Road contractors are required to call the Cable Tube Information centre before they could begin excavation activities, and the centre provides names of utility companies with facilities in the excavation area. The contractor must then contact each utility company, which must provide information on the location of their facilities.

Road contractors in England also must also notify all affected utilities before they could begin to excavate. England is now creating a computerised registry of all utility installations. Despite these efforts, damage to underground utilities continues to occur. Extensive programmes have been developed in the United States to reduce damage to underground utilities caused by excavation activities. One-call notification centres have been established in every state. Contractors are required by law to call these centres and provide appropriate information before they excavate. The centres must then notify all affected utilities. When notified, utilities must visit the proposed excavation site within specified time and mark the location of their facilities with paint or flags. After calling notification centres, contractors must wait for the site to be marked, protect markings after they are placed, and hand dig within a specified distance about half a metre on either side of marked lines.

• **Minimising pavement cuts**

Pavement cuts are a significant problem in Europe. Motorways and most other trunk roads are protected from road openings made by utilities, but often lower-volume roads and streets are not. In Germany, underground utility crossings on trunk roads are made by boring, jacking, directional drilling, or similar means. Pavements are cut for utility crossings on lesser roads and to accommodate fibre optics on streets, a trend that is becoming of concern to road officials. Pavement cuts are also a problem in the United States. Pavements on low-volume, low-speed road are cut for utility crossings routinely in rural areas. In cities, pavements frequently are cut to access the many utilities located longitudinally beneath streets. Fibre optics installations are becoming particularly troublesome as streets are torn up for installations and then poorly repaired. Even excellent repairs are not sufficient to prevent reduced pavement life in cut pavements. The scanning team recommends that the United States makes a greater effort to use trenchless technologies for road and street crossings, and to control the frequency of pavement cuts to access or install utilities under city streets.

• **Multidisciplinary team approach**

Several countries employ a project management approach, including the use of multidisciplinary project teams. Teams in the Netherlands are responsible for a project from planning through construction. Other Dutch management practices include:
   o   Road corridor and utilities' participation from the planning stage;
   o   Budget and schedule commitments with sign-off by functional representatives and managers;
   o   Treating road corridor activities as a critical path element of project management; and
   o   Accountability for delivery on commitments.

In some countries, such as England, the project team is in a separate part of the transportation agency. The British use a framework document to facilitate coordination and communication by defining the respective roles and responsibilities of lands acquisition personnel and project team members.

Benefits cited as a result of using a project management approach include a shift in employee loyalty from functional units to the project as a whole, better communication and coordination among disciplines, more realistic scheduling, and earlier problem identification and solution.

- **Design-Build**

England uses design-build contracting extensively in its programme to reduce the time required for project development. While contracts include utilities coordination, road corridor acquisition remains with the road agency. Germany established an agency (the German Unity Planning and Construction Company for Trunk Roads, known as DEGES) to expedite new construction and deal with rehabilitation projects important to the reunification of the country. The responsibilities of DEGES include land acquisition on behalf of the transport agency. These practices suggest transportation agencies in the United States may benefit from considering further innovations in the area of design-build contracting, including expanding the design-build contract scope to include some or all road corridor services.

- **Multidimensional and inclusive planning processes**

In several countries, zoning and land use plans prepared at the local or regional level govern decisions about the location of transportation infrastructure. Transportation agencies normally have a role in developing the plans. Redevelopment and transportation infrastructure issues are considered at the same time. Land use planning and modal integration are the major focal points for the transportation agencies and others. The processes in each nation provide for significant input from affected property owners, community members, and local authorities. The planning efforts lead to adoption of a detailed definition of the project before involuntary road corridor acquisitions begin. Norway's planning system operates at country, municipal, and zone levels under its Planning and Building Act. The zone plan is the most detailed and complete. Road decisions are normally made based on municipal and sometimes country level planning. The planning process includes participation from an array of interested parties, including landowners. Germany uses a plan settlement and approval process when projects may significantly affect private parties and there is opposition to the proposed infrastructure. The process includes a public hearing before an independent authority that balances public and private interests, including the needs of utilities. The specific procedures depend on the scope of the project. Two simplified processes are used when the transportation proposal itself is insignificant or there are no significant impacts or opposition, property owners have agreed to the necessary acquisitions, and consensus has been reached with others on matters of public concern. An approved plan sets the alignment and road corridor for the roads. The Tracewet (Route) Act has defined the decision-making process for projects of national importance in the Netherlands since 1994. Under this act, Rijkswaterstaat assesses and balances economic and environmental issues, includes significant public participation in plan development, and looks for agreement with local authorities on proposed projects. Projects must be consistent with local zoning plans, although such plans are often revised to achieve needed consistency. Amendments to the Route Act in 2000 are expected to permit the route plan to prevail over the local land use plan when local agreement and land use plan revisions are not achieved within a reasonable time. A benefit of the zoning plan requirement is that it prevents inconsistent land development from occurring while zone plan changes are pending and after revisions are adopted. It also fosters public acceptance of the proposed projects. In England, more location and design detail is added at each level of planning, with country-level plans being the most explicit. At that level, the plan typically identifies access points along the major road, or trunk roads. The road agency has the power to direct the regional planning authority to deny or grant access with conditions. No general legal right of access to trunk roads exists in England. A major benefit of strong local planning systems in the countries visited is the broad ability to make thoughtful and comprehensive decisions about future needs, including appropriate land use and transportation infrastructure. The system also improves project quality and public support, and creates the opportunity to save considerable time in the project development process. The success of European practices suggest that re-examination of corridor preservation is warranted in the United States, using the "1990 Report of the AASHTO Task Force on Corridor Preservation" as a starting point. The review should consider how states might benefit from lessons offered by the more holistic European approach to land use, environmental, and transportation planning.

* **Definition of problems and solutions**

To improve project quality and save time, European nations typically use the planning process to define specifically problems a project will address and how it will achieve intended results. The objective of this practice is to prevent scope creep, unnecessary work, and late plan changes. The scanning team found the most rigorous example of this in the Netherlands, where the order for property expropriation includes a description of identified transportation problems and a statement specifying how the project will address those problems.

* **Planning stage feasibility analysis**

Several countries use broad feasibility reviews before road corridor acquisition. Items considered in the Netherlands include land use, environmental effects, financing, and engineering. Germany does a similar review that incorporates a cost-benefit analysis of traffic and safety measures.

* **Land consolidation**

A European concept that caught the attention of the scanning team is the land consolidation process. Norway, Germany, and the Netherlands all use some form of this practice. Land consolidation allows the pooling of fragmented land parcels and redistribution using more economically rational parcel configurations. Distribution is based on value, so owners receive land of the same value as the land they put into the pool. This procedure is used primarily in agricultural areas to reorganise properties where a project has adverse effect because of a new alignment or significant widening. Germany also uses the process to implement a new urban zoning plan and its accompanying transportation infrastructure. In Norway, one owner's request can initiate the process, while in Germany and the Netherlands the majority of affected owners must agree to start consolidation. Although this practice may seem foreign to Americans, European property owners are pleased to have more economic parcels and new roads. The road agency benefits because it can reduce the number of road crossings needed to service parcels separated by the road. Used on a voluntary basis, the land consolidation concept may offer promising opportunities in the United States to improve land use and property operating characteristics after a road project is completed.

* **Realistic road corridor budgets and schedules**

Each of the European countries studied typically allocates enough time and money to permit appropriately timed and scoped acquisitions. Each country has its own version of the British principle: *Put individuals affected by a project in positions in which they neither gain nor lose from the project.* This operating rule leads to an owner-oriented process, including broader use of flexible acquisition benefits and property management practices. One example is the British approach under which strip acquisitions with property effects unacceptable to the owner can be mitigated through full acquisition and resale of the affected property. In all the four countries, high priority is placed on settlement of acquisition cases. A premium is often paid to accomplish settlements. A Rijkswaterstaat representative in the Netherlands referred to this practice as making an investment decision that takes into consideration the various costs and benefits of expediting a settlement. In the overall, the European approach results in better processes and outcomes for property owners than often occur in the United States. Property owners find themselves in at least as good a position as they were before the project and sometimes in a better situation. Settlement rates and abutter satisfaction rates are high, which help to avoid project delays.

- **External communication, coordination, and participation**

Each country visited engages in extensive public coordination, and they each characterise the practice as valuable. The countries consistently encourage owner participation in design issues at early stages of project development. Several countries typically have the project manager or designer and the road corridor team member who conduct field reviews with affected property owners early in project development. In England, designers are available to meet with property owners to discuss issues and impacts once the road agency has determined the preferred route. Before construction begins, the contractor holds a public meeting to discuss the work and its expected impacts.

Project teams in the Netherlands contact potentially affected owners early so their concerns can be addressed as the design is developed. The Netherlands is piloting a decision-making process for use on major projects called interactive planning. The goals of the interactive planning process are to speed up projects, gain more public support for projects, and find more creative transportation solutions through participatory problem-solving sessions. Results suggest that a range of public participation models including information, reaction, research, consulting, participation, and partnership is needed to address the variety of circumstances affecting project development and implementation. Broad public participation benefits the transportation agencies by helping them to identify issues and incorporate needs and solutions into the original project design. This practice avoids many late plan changes and improves relationships with affected property owners, municipalities and other parties.

- **User-Friendly road corridor plans**

Road corridor plans observed in Europe are clearer and easier to interpret than plans in many jurisdictions in the United States. In Norway, for example, zone plans are the basis for acquisition. Norway does also minor acquisitions without plans. The Netherlands negotiates with property owners using a general schematic showing existing and proposed road corridor lines and macro-level design information. Parcel-specific acquisition information is shown on a land interest plan that excludes engineering information. The Netherlands does not require final design to move forward with voluntary acquisitions, but detailed plans are used when involuntary acquisitions are done by expropriation. In England, an engineering schematic, land interest plan, and occasionally a model are used at public hearings. The land interest plan shows only the area to be acquired, field reference numbers, and boundaries of the acquired properties. It is the plan used for property owner negotiations and for recording at the registry of deeds. No longitudinal baseline is used on the plans and no reference points are plotted in the field. A unique GIS centre point is used for the parcels acquired. The road agency uses as-built plans of the road to locate points in the field after a project has been completed. Detailed construction or survey plans may be used in litigation if needed. Countries using land interest plans report that plan simplicity is not only acceptable to owners, but makes it easy to explain acquisitions to them. Negotiators typically use engineering schematics to explain construction impacts. One important element that facilitates use of these simpler plans is the national standardised mapping, land registration, and survey system in each country. Nonetheless, European practices suggest that there are opportunities for cost savings and simplification that could be exploited by those states now using highly detailed and complex road corridor plans for negotiations and acquisition documentation.

2.6.3   Utility Accommodation Practices

Accommodating new utilities within road corridors depends on factors such as age of existing infrastructure, past, current and future land uses, and community aesthetic preferences. Placing new utility infrastructure in crowded road corridor, for example, involves avoiding damage to other utilities unlike construction in new developments. In both conditions, planning, designing and the ability to accurately locate other infrastructure are critical elements in protecting new, existing and future facilities. Placement of facilities is dictated not only by the physical limit of road corridor or easement, but also by rules about minimum burial depth and separation distances. State health rules require that water and sewer lines be separated by a minimum distance and/or elevation difference or placed in impermeable materials.

The current approach is related to the use and occupancy agreements. The written arrangements set forth the terms under which the utility is to cross or otherwise occupy the road corridor with reference to the following:

i)   The TANROADS standards for accommodating utilities which include:
   •   The horizontal and vertical location requirements and clearances for the various types of utilities;
   •   The applicable provisions of government or industry codes required by law or regulation including road design standards or other measures which the agency deems necessary to provide adequate protection to the road, its safe operation, aesthetic quality, and maintenance;
   •   Specifications for and methods of installation; requirements for preservation and restoration of road facilities, appurtenances, and natural features and vegetation on the road corridor; and
   •   Measures necessary to protect traffic and its safe operation during and after installation of facilities, including control-of-access restrictions, provisions for rerouting or detouring traffic, traffic control measures to be employed, procedures for utility traffic control plans, limitations on vehicle parking and materials storage, protection of open excavations, and the like must be provided.

ii)   A general description of the size, type, nature, and extent of the utility facilities being located within the road corridor.
iii)   Adequate drawings or sketches showing the existing and/or proposed location of the utility facilities within the road corridor with respect to the existing and/or planned road improvements, the travel way, the road corridor lines and, where applicable, the control of access lines and approved access points.
iv)   The extent of liability and responsibilities associated with future adjustment of the utilities to accommodate road improvements.
v)   The action to be taken in case of non-compliance with the transportation department's requirements.
vi)   Other provisions as deemed necessary to comply with laws and regulations.

Figure 2.18 shows the utility permit process flow chart as given by Manual on Accommodating Utilities in Pierce County Road Corridor (2004).

Figure 2.18
Utility road corridor permits process flow chart

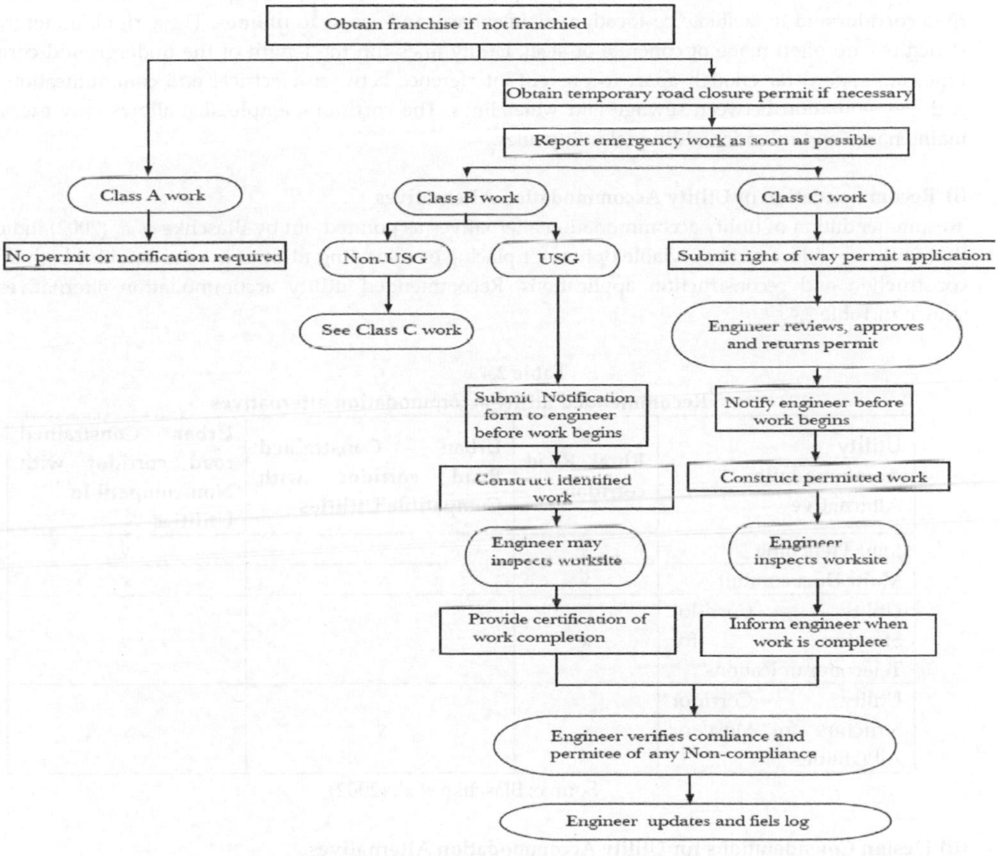

Obtain franchise if not franchised

Obtain temporary road closure permit if necessary

Report emergency work as soon as possible

Class A work | Class B work | Class C work

No permit or notification required

Non-USG | USG

Submit right of way permit application

See Class C work

Engineer reviews, approves and returns permit

Submit Notification form to engineer before work begins

Notify engineer before work begins

Construct identified work

Construct permitted work

Engineer may inspects worksite

Engineer inspects worksite

Provide certification of work completion

Inform engineer when work is complete

Engineer verifies comliance and permitee of any Non-compliance

Engineer updates and fiels log

Source: Manual on Accommodating Utilities in Pierce Country Road Corridor (2004).

2.6.4 Emerging Practices

One of the major challenges has always been the limited space for utilities. Thus, one pertinent question is: What is the most effective way to manage road corridor space so that an agency can allow all of the utilities required? And how can it be made easy for utilities to access and maintain their lines? Answers to these questions were provided through a research project by Texas Transportation Institute (TTI). Documented in utility corridor structures and other utility accommodation alternatives in TxDOT road corridor Kuhn *et al.* 2003), the research demonstrates that joint trenching (parallel lines along the same trench) and multi-duct conduits (tubes designed to hold multiple lines) are feasible alternatives for TxDOT as it works toward more effective road corridor management. TTI also explored utility corridors as one way to accommodate utilities in the limited road corridor space. A utility corridor is defined as an area of road corridor designated or used for the joint location of utilities, either public or private. Similarly, a public utility corridor is a corridor for the placement of water, wastewater, electric and gas system infrastructure for which easements have been granted, that may also provide open space functions. A utility corridor structure is a joint-use facility or conduit constructed or installed within a road corridor that can accommodate a variety of utilities to minimise congestion of utilities within the road corridor and to facilitate co-location, maintenance, and access to utilities. These rigid, underground structures are often made of concrete or steel. Utility lines run the length of the underground corridor. Lines are spaced far enough apart to prevent interference between electrical and communication lines and contamination between sewage and water lines. The corridor's ample size allows easy access for maintenance and room for additional utility lines.

(i) Recommendation of Utility Accommodation Alternatives

Recommendation of utility accommodation alternatives as pointed out by Blaschke *et al.* (2002) indicated that utility corridors can be a viable option for placing or relocating utilities in road corridor both in new construction and reconstruction applications. Recommended utility accommodation alternatives are shown in Table 2.4.

Table 2.4
Recommended utility accommodation alternatives

Utility Accommodation Alternative	Rural Road corridor	Urban Constrained Road corridor with Compatible Utilities	Urban Constrained road corridor with Non-compatible Utilities
Joint Trenching	√	√	√
Multi-Duct conduit	√	√	x
Utility Corridor Structure for Telecommunications	x	√	√
Utility Corridor Structure for Most or All Utilities	x	√	x

Source: Blaschke *et al.* (2002)

(ii) Design Considerations for Utility Accommodation Alternatives.

Utilities in the road corridor can be differently accommodated relative to the available space and costs. Utility corridor structures with and without walkway in Figure 2.19 is one of the design considerations for utility accommodation alternatives.

Figure 2.19
Examples of utility corridor structures with and without walkway.

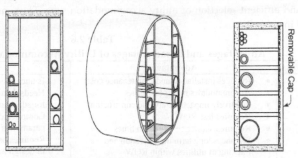

Source: Blaschke *et al.* (2002)

Table 2.5 shows utilities accommodation alternatives with their related considerations.

Table 2.5
Design Considerations for Utility Accommodation Alternatives

Utilities Accommodation	Design Consideration
Joint Trenching	• The minimum separation (S) described by codes, ordinances, and utility power must be maintained. Minimum depths (D) must meet policies of TxDOT and must comply with codes, ordinances, and utility owner policies • Underground detectable warning tape shall be placed in all trenches at one foot below grade
Multi-Duct Conduct	• The compatibility of utilities must be determined prior to placement within a multi-duct conduct
Utility corridor Structure	• Utility corridor structure may be designed as large structures that provide a corridor as a walkway throughout the facility or as a smaller structure without a walkway and with accessibility provide at designated intervals
Telecommunications and compartible utilities	• The design of the structure should be based on the height
	• For large structures, extra reinforcing steel may be required as the structure serves as a retaining wall
	• Access point should be provided at 500 to 1000 ft intervals, depending upon the type of utilities to be installed
	• The profile of the structure should facilitate drainage. Access should be provided and a drain should be installed where practical
Most of all utilities	• Hangers should be designed and spaced per the manufacture's specifications for the types of utility line the hanger will support
	• The structure should be water proofed as per TxDOT specification.
	• The utility company must provide illumination and certification in accordance with Federal, State, and local regulation and ordinances

Source: Blaschke *et al.* (2002)

(iii) Advantages and Disadvantages of Utility Accommodation Alternatives
The utility accommodation alternatives have advantages and disadvantages, which can influence the effective and efficient selection of utility accommodation type. These advantages and disadvantages are illustrated in Table 2.6.

Table 2.6
Advantages and Disadvantages of Utility Accommodation Alternatives

Alternatives	Advantages	Disadvantages
Joint Trenching	• Reduces installation and maintenance costs • Accommodates multiple utilities • Positively impacts safety and construction • Requires less ROW • Requires shorter construction times • Enhances long term identification and tracking of utilities within ROW • Minimises impact on the environment • Provides benefits in areas where the type of soil involves expensive excavation costs	• Is uncommon in underground facilities • Needs detailed coordination between utilities for successful compilation • Complicates agreements for design parameters and shared cost • Requires one utility to take a leadership role in design and construction
Multi-Duct Conduct	• Accommodates multiple utilities at less cost than multiple installations • Plans and installs future growth at minimum cost • Requires less ROW • Positively impacts safety and construction • May allow trench less boring through installation techniques • Heavily used in underground installation • Requires shorter construction times	• Is feasible only for companies utilities • May be difficult to estimate size for future growth • Needs detailed coordination between utilities for successful compilation • Complicates agreements for design parameters and shared costs • Requires one utility to take a leadership role in design and construction
Utility corridor Structure 1) Telecommunications and compatible utilities 2) Most of all utilities	• Many minimise total road corridor necessary • Assures no locations for all telecommunication facilities • Reduces overall construction and installation time • Enables planning for significant future growth by all utilities • Reduces repair time in the event of a break or malfunction	• May be considerably more expensive than joint trenching or multi-duct conduits • Large structures may require more ROW than other methods • Requires designs to include additional items not typically addressed in utility installations such as lighting ventilation and drainage • Requires specific spacing location and casing requirements for non-compatible utilities • Needs detailed coordination between the utilities for successful completion
	• Minimises impact from adjacent construction activities	• Complicates agreement for designing parameters and shared costs • Requires one utility to take a leadership role in design and construction • Requires long-term strategies for maintenance and repair procedures • May create significant security concerns

Source: Blaschke *et al.* (2002)

2.6.5   Transportation Utility Corridor

The use of Transportation and Utility Corridors (TUCs) is an alternative approach. In Alberta, Canada, the Government Organisation Act enabled the establishment of Restricted Development Areas (RDAs) to coordinate and regulate the development and use of certain areas. The Calgary RDAs and the Edmonton RDAs are of particular interest because of the designation of TUCs within those RDAs. The TUCs were established on the principle that long-term planning for the accommodation of a number of transportation and utility facilities within a TUC can maximise its use. The TUCs protect ring road and utility alignments from advancing urban development. Specific advantages to the use of TUCs include land conservation, limited environmental disruption, administrative efficiency, safety, land use certainty, assured alignments for future users, and open space use.

The Government of Alberta in the US in the mid 1970s established Restricted Development Areas (RDAs) around the cities of Calgary and Edmonton. The lands included in these RDAs were designated for the Transportation and Utility Corridor (TUC) uses, being the high standard ring road systems, major power lines, pipelines and linear municipal utilities (as shown in Figure 2.20). The TUCs were seen as the most effective means of providing long-term alignment for future ring roads and major linear utilities needed to serve these expanding urban areas. There are four major types of utility components within the TUCs:
- Ring Road (freeway) and buffer to allow for future widening and/or realignment
- Pipelines
- Power lines – 69 kV and above
- Municipal services (power lines less than 69kV, storm water management facilities associated with the ring road and regional water, sanitary or storm trunk sewers)

In addition, an access component is designed to maintain access for compatible secondary uses, for the maintenance of the existing utilities, and for the installation of primary use facilities.

Figure 2.20
Typical TUC cross-sections.

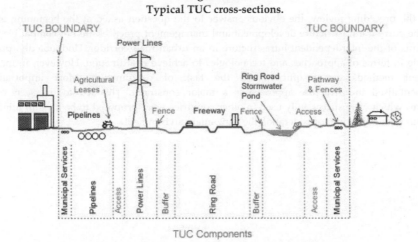

Source: Ministry of Infrastructure-Alberta (2007).

2.7   Conclusion

A number of theoretical models exist to conceptualize characteristics of infrastructure interdependencies and their impact as cited by Rinaldi *et al.* (2001). Empirical work has been less common, and anecdotal evidence tends to be used more commonly as the basis for describing and modelling the behaviour of interdependent infrastructure. Identifying, understanding and analysing the interdependencies among infrastructures have taken on increasing importance during the past decade. Interdependencies, however, are a complex and difficult problem to analyse as shown in this chapter.

The 'science' of infrastructure interdependencies is relatively new and current modelling and simulation tools are only beginning to address the issues. Many models and computer simulations exist for aspects of individual infrastructures (e.g. load flow and stability programmes for electric power networks, connectivity and hydraulic analysis for pipeline systems, traffic management models for transportation networks), but simulation frameworks that allow the coupling of multiple, interdependent infrastructures are only beginning to emerge.

In order to clarify better the issue of geographical interdependency of civil infrastructure systems, Akinyemi (2007) has defined local events that affect the infrastructure systems in an urban road corridor and the effects they have as follows:

* The location of a set of elements of "A" has an effect on the structural and/or operational performance of "B" and vice-versa;
* The process of construction of a set of elements of "A" affects the structural and /or operations performance of "B" and vice-versa;
* A set of maintenance repair and renewal activities of "A" affects the structural and /or operations performance of "B" and vice-versa;
* A fault or problem in "A" leads to, contributes to, or magnifies a fault or problem in "B" and vice versa.

Based on the preceding review, the obvious answer to the question asked at the beginning is that, in general, the current infrastructure development and management practices are not compatible with the requirements of the interdependent infrastructure in an urban road corridor. Undoubtedly, progress is being made in terms of approaches and technologies to achieve coordination. However, in many cases the current methods are insufficient for the task of coordination. Most importantly, the compartmentalised infrastructure approach is a major constraint. This led to overseen problems. Alternatives which were successfully used in other countries were proposed to be used to minimise the conflicts but after taking into consideration the Tanzanian environment.

3

3.0   RESEARCH METHODOLOGY

This chapter presents the research methodology under the following sections: Research Design and Approach; Selection of the Case Study Area; Data Collection Technique and Procedures and Data Analysis Methods.

3.1   Research Design and Approach

The study adopted a combination of quantitative and qualitative research approaches although the former dominated. The combined approaches aimed at increasing reliability of information, thereby making qualitative complement the quantitative information.

In order to provide answers to the research questions and to test the hypotheses, the current state of knowledge on linkages among infrastructure assets in the road corridor were reviewed. In addition, a review was conducted to study the current state of practices on development and management of infrastructure assets and road corridor in Dar es Salaam.

3.2   Selection of the Case Study Area

Dar es Salaam is the major commercial city in Tanzania. Dar es Salaam was selected because it had most of the infrastructure interactions required to attain the objectives. Given the fast growing population, the road corridor is being threatened by emerging human related activities coupled with faulty urban infrastructure interaction thereby affecting movement and safety.

Dar es Salaam city is constituted of a radial system of roads. Major highways and arterials of either 4 lanes-2 way or 2 lanes-2 way converge in the city centre where most of the traffic ends and originates. Collector and local streets connect to these roads to form the city's road network. Most of the road links in the corridors are characterised by the kerbs that are common in downtown local streets but rare in major arterials that shoulder most potholes and are occupied by various activities. There are pavements that exist on many major arterials while medians exist on all multilane facilities.

Urban roads in Dar es Salaam are functionally classified as arterial, collector, or local streets (JICA, 1990). Arterial roads are designated for major traffic movements with high volumes and high design speed. Collectors are designated for reduced movement function and may be either primary or secondary. Local roads are designated primarily for accessibility. Most arterial roads are four-lane two-way facilities with medians separating the two directions of traffic travel and few are two-lane two-ways with no medians. All collector and local roads are two-lane two-way facilities. The common lane width for arterials is between 3.5 – 3.7 m and for collector and a local road is between 3.0 – 3.5 m. Most of arterial and collector roads have shoulders, some of which are unpaved. The width of paved shoulders ranges between 1.0 –

2.0 m. Speed limits for major arterials range from 50 to 80 kilometres per hour (km/h) depending on the location of the road within the city area.

3.2.1   Site Selection

Traffic and road corridor conditions were identified, and specific arterial roads and sites were selected. The selection was limited by time and budgetary constraints. Selection was deemed important because results were expected to represent the whole study area of Dar es Salaam. Road corridors were selected based on the presence of a wide range of urban infrastructure and traffic flow conditions. Also they were selected on the basis of their physical and geometric quality that could support this kind of study.

A reconnaissance survey was made throughout the main roads in the city and suitable road sites of the study were identified. When selecting the sites, the roadway condition, road side infrastructure, underground infrastructure environment, and interaction of the infrastructure in the road corridor were the main factors taken into account. The study covered corridors of major arterial roads in Dar es Salaam, namely, Morogoro, Mandela, Kilwa, Bagamoyo and Nyerere. Table 3.1 shows the five selected principal arterial roads demarcated to form a total of 70 kilometres.

Table 3.1
Selected urban principal arterial roads

No	Name of the Road	Total length (km) of the road	Length (km) covered under the study	Remarks
1	Morogoro Road	20	18	North/West Corridor
2	Bagamoyo Road	30	17	North Corridor
3	Nyerere Road	15	6.7	To International Airport
4	Kilwa Road	12	6.5	South Corridor
5	Mandela Road	21.8	21.8	Ring Road
	Total	96.8	70	

Source: TANROADS (2007)

i)   Morogoro Road Corridor

The road section under the study is demarcated from Kilwa Road to Kimara (Mbezi) site, 18 km from the city centre, has 2 lanes 2 way undivided west of Nyerere Road and 4 lanes 2 way divided road east of Nyerere Road.

Morogoro Road is a North-West dual carriageway road that connects Dar es Salaam and central regions of Tanzania and nearby countries. The road is extremely important for the regional economy since most freight is shipped into and out of the Dar es Salaam port through the route. It also connects Tanzania and landlocked countries of Uganda, Rwanda, Burundi and the Democratic Republic of Congo (DRC); these use the road for transporting goods to and from Dar es Salaam port.

ii)   Bagamoyo Road Corridor

The road section is demarcated from Kilwa Road to Mwenge, 17 km from the city centre, 2 lanes 2 ways undivided sections along Ohio Street, 4 lanes 2 ways divided sections along Alli Hassan Mwinyi, and 2 lanes 2 ways undivided sections along new Bagamoyo Road.

The road connects Dar es Salaam to the Northern Coastal Regions and is an important mobility route to the residents of newly developed suburban areas along the axis. Located west of Kawawa Road, it consists of one lane per direction with no median separation and from the east there are two lanes per direction (with raised medians).

iii)   Nyerere Road

The road is demarcated from Tanganyika Motors, a junction with Bagamoyo Road to Mwalimu Nyerere International Airport, with 4 lanes 2 ways divided road section. Nyerere Road functions as a vital industrial road in the city with several manufacturing and distributing industries located along it. Also the road provides access to Mwalimu Nyerere International Airport. It is a high standard dual carriageway with pedestrian walkways on both sides. The main traffic along the corridors consists of trucks from industries and private vehicles to and from the international airport. The 6.7 kilometres of this road, which lie within the Central Business District (CBD), was studied.

iv)   Kilwa Road

The road section is demarcated from Kivukoni to Sabasaba grounds 12 km from the city centre. It has 2 lanes-2 way undivided road section. The road connects Dar es Salaam to coastal regions (in Southern Tanzania) and is the only route which links Kigamboni peninsula with the mainland. Currently it consists of only one lane per direction. The stretch of 6.5 kilometers of the road that lie within the Central Business District (CBD) in the Dar es Salaam city was studied. This route is dominated by the commuter buses operating in the city commonly known as daladala. Many people living along the route are middle and lower class; therefore, many people are used to walk along the road section.

v)   Mandela Road

The road section is demarcated from Bagamoyo Road to Bandari along Kilwa Road; having 4 lanes 2 ways divided road section. Mandela Road is the ring road joining the above-mentioned arterial roads in Dar es Salaam city. The road serves as an expressway from Dar es Salaam port. The whole of 21.8 km of the road lies within the CBD in Dar es Salaam. Most traffic along the route comprises cargo trucks from the port to the countryside or nearby countries.

3.2.2   Location of the Study Area

Tanzania has an area of 945 square kilometres (km²) with a population growth from 23.1 million in 1988 to 34.4 million in 2002 (Population Census 2002). The population projections show that in the year 2006 Tanzania has a population of 37.9 million; this is according to the National Population Policy of the United Republic of Tanzania (2006). Figure 3.1 shows the road network of Tanzania in general and the study area, Dar es Salaam city, in particular.

Figure 3.1
Map of Tanzania and Its Road Network

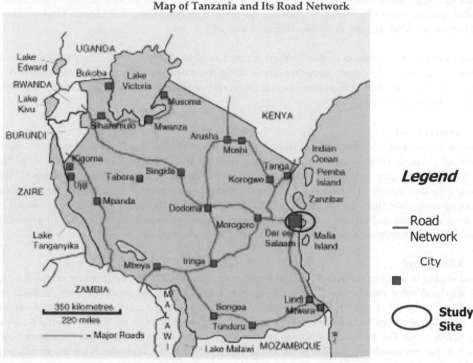

Source: Ministry of Infrastructure (2006)

Dar es Salaam lies between 6.34' and 7.10' southwest of the Indian Ocean. The city is potentially the engine of socio-economic development in Tanzania by virtue of being the main administrative, industrial, commercial, trading, educational, and cultural and transportation centre in the country and to nearby countries. It is the largest urban centre in Tanzania with an area of 1,393 km². Between 1967 and 2002 the population increased from 843,090 to 2,497,940 (Population Census 2002). The city's population is growing at an average rate of 8 percent per annum with spatial expansion of about 7 percent per year and is one of the fastest growing cities in Sub-Saharan Africa (SUDP, 2002). Dar es Salaam is of interest as it has experienced rapid urbanisation and the highest number of road traffic accidents in the country.

The city has a total road network of 1460 km of arterial, regional and local roads. Four of its main roads are radial and carry most of the vehicular traffic entering and leaving the city. Another one is a semicircular ring road connecting the four radial roads with Dar es Salaam harbour. The city has three municipalities namely Kinondoni (where the Bagamoyo road traverses), Ilala (where the Morogoro and Mandela roads traverse) and Temeke (where the Nyerere, Mandela and Kilwa roads traverse, as illustrated in Figure 3.2).

Figure 3.2
Map of Dar es Salaam urban roads

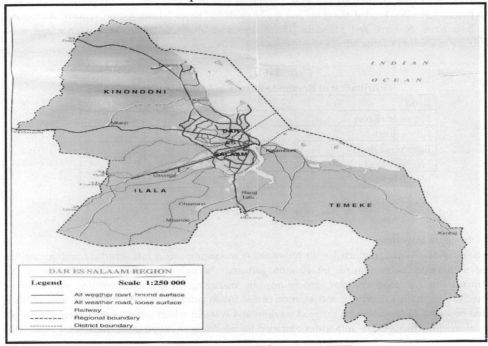

Source: Ministry of Infrastructure (2007).

3.3 Data Collection Techniques and Procedures

To attain the objectives of the study three categories of data, namely infrastructure data, traffic data, and accident data needed to be collected. Consideration of these variables for data collection was partly guided by the site selected. At the same time the type of variables to consider was guided by the objectives of the study. For the purpose of data collection, it was important to identify the criterion and predictor variables that were most relevant for this study. The availability and the details of these data affected the results of the study.

3.3.1 Data

Primary data were collected by use of structured questionnaire, in-depth interviews, road inventory survey, documentary review, and surveillance cameras.

i)   Questionnaire
The questionnaire was used to collect information from different users of the selected roads (see Annex I). It consisted of mostly close-ended questions, which aimed at gathering information on interaction of urban infrastructure, movement, road surface condition and safety of movement. It also included travel characteristics, infrastructure conditions, infrastructure operations in the road corridor and the observed effects.

The questionnaires were distributed to a total of 500 respondents, comprising 270 (54 percent) males and 230 (46 percent) females, located along the five major roads in Dar es Salaam region, namely, Morogoro, Kilwa, Bagamoyo, Nyerere and Mandela. The distribution of respondents per residence and road locations are as shown in Table 3.2.

Table 3.2
Distribution of Respondents by Road Location (N=500)

Road	N	%
Morogoro Road	100	20
Kilwa Road	100	20
Bagamoyo Road	100	20
Nyerere Road	100	20
Mandela Road	100	20
Total	500	100.0

Source: Field Data Survey (2007)

ii)   In-depth Interviews

In-depth interviews were conducted with stakeholders managing urban infrastructure and authorities dealing with safety of movement. Information gathered through interviews included state of the infrastructure in the road corridor, procedures for managing urban infrastructure, operation of infrastructure in the road corridor and location of the urban infrastructure in the study area. Also it included problems faced in sharing the road corridor and ways in which safety of movement is affected by the urban infrastructure. The authorities contacted for interview are listed in the Table 3.3.

Table 3.3
Authorities involved in the In-depth Interviews

Authority	Acronym	Responsibility
Ministry of Infrastructure Development	MOID	Road management and policy making
Tanzania National Roads Agency	TANROADS	Managing trunk and regional roads
Tanzania Telecommunication Company Limited	TTCL	Managing telecommunications
Tanzania Electricity Supply Company	TANESCO	Managing and supplying electric power
Dar es Salaam Water Supply Authority	DAWASA	Water supply management
Dar es salaam Sewerage Management Company	DAWASCO	Managing sewerage
Ministry of Home Affairs	MOH	Traffic Police for recording traffic accidents and road law enforcement
Dar es Salaam City Council	DCC	Managing urban infrastructure in the city
Municipal Council	MC	Managing infrastructure in the municipals
Survey and Mapping Division	SMD	Surveying and Producing of maps

Source: Field Data Survey (2007)

iii) Surveillance Cameras

Surveillance cameras were used to observe and capture activities as well as operations taking place in the study area. Also observed patterns included adherence to traffic laws and regulations by road users. Observed incidences on the road corridors included damaged road surface due to faulty interaction of urban infrastructure, and interaction between road corridor environment, and traffic movements. The observation was done for 24 and 8 hours using fixed surveillance cameras and mobile cameras respectively.

Full-scale field data collection was conducted along five selected road corridors. The fixed surveillance cameras are shown in Figure 3.3 and time schedule of the exercise is shown in Table 3.4. Data from the field were obtained from 10 sites at an average of 24 hours, surveillance camera data were collected for one year.

- **Fixed surveillance cameras**

Surveillance cameras were fixed in various sites as elaborated in Table 3.4 and shown in Figure 3.3. The Video Casette Recorders (VCRs) on the sites were fixed in two boxes one made of aluminium and the other made of iron bars in order to prevent vandalism. The six surveillance cameras were fixed at different selected sites by using mobile bucket vehicles, which helped in fixing them at different desired heights in order to precisely capture on time observations.

Table 3.4
Sites of Surveillance Cameras

S/N	Road name	Surveillance camera site 1	Surveillance camera site 2
1	Morogoro	NSSF (Morogoro/Nyerere) junction	Ubungo Junction (Morogoro/Mandela)
2	Bagamoyo	Salender (Alli Hassani Mwinyi/Kenyata) junction	Mwenge (Bagamoyo/Mandela) junction
3	Nyerere	TAZARA (Nyerere/Mandela junction)	
4	Kilwa	Railway Station	DSA at Kilwa and Mandela roads junction
5	Mandela	Chang'ombe- Police junction	TAZARA

Source: Field Data Survey (2007)

- **Mobile surveillance cameras**

Two surveillance cameras were fixed in a mobile motor vehicle at either side for onsite investigations. One camera recorded features on the left side, while the other recorded features on the other side.

On inspection, video recording of 70 km selected road network was undertaken. It was followed by video rating that was done in the laboratory, of the required attributes.

Different methods were applied for different data types involved in the collection scheme. Surveillance camera recording was chosen as the primary method of data capture during observations and operations. Through such cameras the road corridor condition, traffic flow characteristics, which included traffic flow, composition, speed, traffic behaviour due to operation of urban infrastructure and safety of movement of people and vehicles in the road corridor could be recorded. Effects of urban infrastructure on each other and on movement as well as safety were also recorded. Data collection focused on the following components: urban infrastructure interaction, road corridor condition, activity of urban infrastructure implemented in the road corridor, traffic flow, speed, traffic composition, traffic volume and affected movement including safety.

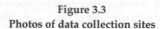

Figure 3.3
Photos of data collection sites

Source: Photo Taken at the Surveillances Camera Sites (2007)

iv)   Road Corridor Inventory Survey

An urban infrastructure inventory survey was conducted along Morogoro, Bagamoyo, Nyerere, Kilwa and Mandela roads. All road corridors along these routes were surveyed between 8 a.m and 6 p.m on a typical weekday.

In total, 91 km of road sections in the study area were surveyed. Survey data contain details of existing urban infrastructure found in the road corridors, distance of each infrastructure from the road surface, their condition and the land use characteristics. The urban infrastructure and features found in the road corridor were recorded in a form prepared for data collection. The form is appended in this report. The road corridor inventory concerned the items described below.

- **Road network**

There are five roads under the study area, which are the major arterial roads in Dar es Salaam. As mentioned earlier, the roads are Morogoro, Nelson Mandela, Kilwa, Ali Hassani Mwinyi and Nyerere. The roads were divided in the link road sections and junction for close investigation. A total of 48 km link study sites were selected on these main roads, which represent various geometric and traffic conditions in the city. They included 18 km undivided road sections and 30 km divided road sections. The length of link sections varied between 0.40 km and 3.21 km with an average of 0.92 km. In summation, the total length of road sites came to 91 km with clear demarcations of the start and the end of selected roads under the study. The road sections were defined by, start and end-points, with a Global Positioning System (GPS) device. The road sections were divided into sub-sections by using waypoints. The waypoint positions were generally supplied with a GPS reference. The infrastructure and features on each link were recorded differently on a special data sheet.

Given limited information available on the geometry and location of infrastructure, actual measurements and visual inspections were made so as to collect necessary infrastructure data along each road site in Dar es Salaam. Geometric characteristics were measured by using a tape measure. Recorded measurements were road width, lane width, shoulder width and study segment lengths. Actual measured values and visual inspections were recorded in data collection forms prepared for the link and junction sites separately, including road inventory in the study area. The data collection form is appended as Annex I.

All activities conducted on the road were observed for the whole period of the research. Observed activities included repair and rehabilitation works of the road sections. Observed were the effects of repair and rehabilitation to underground pipes including cables, roadside infrastructure and other features located in the road corridors. Observed also were the drainage system types, condition, their effects on infrastructure and movement of traffic for the entire period of research. Water on the road surface and behaviour of underground infrastructure to the road surface were observed. They were leakages of water and sewer pipes, burst of sewer and water pipes and their effects to the road surface. Other types of infrastructure elements related to safety were also observed. Cuts of road sections for underground infrastructure repair and rehabilitation works were also observed. Monitoring of the cut sections was done to find out how the cuts were done, how they were reinstated and how the section failed. Change of behaviour of traffic movement during underground infrastructure work and its effect to movement and safety were also monitored.

- **Underground infrastructure**

Present underground infrastructure in the road corridor was also assessed during surface investigation through manholes placed on the carriageway. Recorded existence of underground infrastructure included number of manholes, their types and condition. Presence of underground infrastructure was also observed through side connection of the network and through leakages on road surfaces. Recorded for underground infrastructure were water and sewer pipes, electric and telecommunication cables as well as oil and gas pipes. These infrastructures were further observed at an area where excavations and repair were being undertaken.

- **Roadside infrastructure**

Urban infrastructure along the roadside was easily observed since it was above the ground. Recorded, were the number of poles, water pipes, sewer pipes and cables for electric power as well as telecommunications. Distance of poles from the road edge, their condition and land use characteristics along the routes, were recorded.

Figure 3.4
Most common road corridor features in the study area

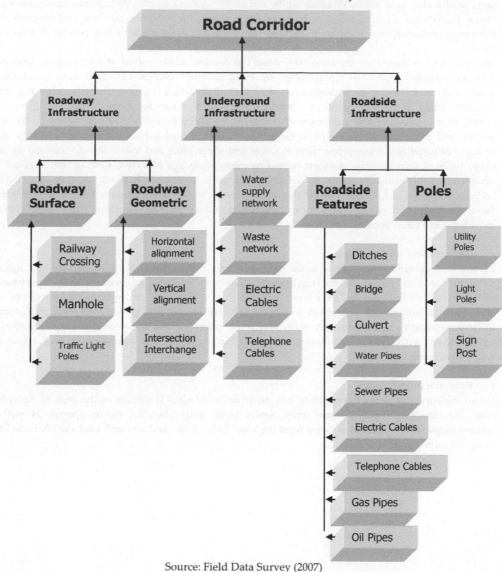

Source: Field Data Survey (2007)

Figure 3.4 shows commonly found features on roadway, underground and roadside infrastructure in the road corridor.

3.3.2   Secondary Data

Collected secondary data included road traffic accident data obtained from the Ministry of Home Affairs, and road network data obtained from the Ministry of Infrastructure Development. The data on urban roads were obtained from Dar es Salaam City Council and Municipal Councils. Data on the telecommunication network were obtained from Tanzania Telecommunication Company Limited, (TTCL) and electricity network data were obtained from Tanzania Electric Supply Company, (TANESCO).

i)   Road Traffic Accident Data
Data on road traffic accidents were collected from the Traffic Police. Accidents were recorded from the accident report book. Accident data included date and time when a person was involved in the accident; the severity of accidents, (whether the accident was fatal, serious or involved minor injuries), location of accidents and details of vehicles involved; loading details and vehicle defects; nature of accidents; for example, details of collision, and pedestrians or objects involved, damage, injuries and casualties. Road traffic accident data were considered important because the study required observing the relationship between urban infrastructure and safety of movement. Accidents are indicators of safety of movement.

ii)   Vehicle Statistics Report
Vehicle statistics were collected from the Ministry of Infrastructure Development with the aim of finding out vehicle importation rate and traffic flow in the study area. According to the 2000 report of the Ministry of Works, estimated annual growth rate of vehicles operating in Dar es Salaam is 4 percent. The traffic count was conducted at selected positions in the study area. Traffic count was considered important because the study required to establish the extent of conflict between urban infrastructure and traffic movement.

iii)   Documentary
Supporting data and information were obtained from the authorities managing urban infrastructure and road safety. They included the Ministry of Home Affairs, (MoH), Traffic Police, the Ministry of Infrastructure Development (MID), Tanzania National Roads Agency, (TANROADS), Tanzania Telecommunication Company Limited, (TTCL), Tanzania Electric Supply Company, (TANESCO), Dar es Salaam Water Supply Authority, (DAWASA) and Dar es Salaam Sewerage Management Company, (DAWASCO). Data from these organisations consisted of reports and the drawings of their network in the road corridor.

3.4   Data Analysis Plan

3.4.1   Data Processing

Data processing essentially concerned registering of relevant events not directly measured in the field and creating merged files with all relevant data for each site. This was finally used to obtain incidences, traffic flow, speed, vehicle types and effect of the urban infrastructure on safety of movement. In this case, events recorded in the video from the field were retrieved from the VCR and processed to add attributes which had not been noted on the site.

Figure 3.5
Video rating data processing

Source: Field Data Survey (2007)

Speed is one of the variables that was not directly measured in the field but was obtained from the recorded video. In essence, it was obtained through a reduction process. Data process involved the use of laboratory equipment, which included two television-monitors, two Video Cassette Recorders (VCR) with push buttons to freeze and run the film, paper and pencil. The synchronised video recordings from each end of the long base were shown simultaneously in the laboratory on two adjacent video monitors, with the field survey clock times displayed as shown in Figure 3.5.

 Data reduction procedure for traffic flow and composition involved simple counting of vehicles from video images in the television-monitors as they passed the studied section. Its composition was equally identified from the images. Data for incidents were recorded and later on transcribed into a computer file.

3.4.2   Data Analysis Methods

Four approaches were used in analysing the collected data. Quantitative data recorded in nominal and ordinal way were analysed by using the SPSS. Surveillance camera data and road corridor condition survey were analysed by using Road Corridor Safety Analysis (RCSA) system (see chapter 8). The data collected from the in-depth interview of urban infrastructure stakeholders were analysed qualitatively by comparing the perception of the majority of respondents. Road condition survey and traffic were analysed by using Road Mentor Database – RMMS based programme of the Tanzania National Roads Agency (TANROADS). Lastly, the road traffic accident data were analysed by using Micro Accident Analysis programme (MAAP) for windows.

PART TWO: ANALYSIS OF INTERACTION OF INFRASTRUCTURE

Part two of this thesis focuses on analysis of interaction of infrastructure in the road corridor and the extent to which such interactions significantly contribute to impairing road corridor environment. It also presents the relationships between the road corridor environment on the one side and movement and safety on the other side. This part is divided into two chapters. Chapter 4 provides interaction of infrastructure and their effects on road corridor environment in Dar es Salaam as well as an approach to the development and management of infrastructure systems in the road corridor. Then, chapter 5 presents how infrastructure interaction in the road corridor affected the characteristics of movement and safety.

4

4.0   INTERACTION OF INFRASTRUCTURE IN THE ROAD CORRIDORS IN DAR ES SALAAM

This chapter addresses the interaction of urban infrastructure in the road corridors in Dar es Salaam as a typical city in a developing country. It presents the extent to which such interactions significantly contribute to impairing infrastructure operations and how they affect the condition of each other within the road corridors. The chapter is especially concerned with the way elements that constituted the road corridor related to each other and the extent to which such relationships impacted on infrastructure operations. Furthermore, this chapter presents the way incumbent infrastructure development and management approaches contributed to creating and escalating the observed degradation of service delivery.

4.1   Infrastructure Layout and Physical Condition

Urban infrastructure is regarded to be a set of physical assets needed to supply certain desired services, link people with services and resources as well as support community goals. They are a complex web of public and private assets. They are created and operated within layers of government that have varying jurisdiction over their locations, design, pricing, accessibility and general operation.

In understanding the physical conditions of urban infrastructure, collected data from inventory survey in the road corridors under the study were assessed to establish whether or not their layout exhibited any relationship and also to understand their conditions. Issues of particular focus were locality of urban infrastructure facilities, facilities network crossing topologies and physical conditions of the facilities.

An inventory of the existing urban infrastructure in road corridors of the study area was carried out in the field. The exercise started by identification of infrastructure facilities that existed within the study corridor. Furthermore, it involved review of corresponding data from urban infrastructure inventory obtained from infrastructure operators within the corridors.

Figure 4.1 shows the urban infrastructure assets, which were found to exist within the study area. They comprised of multiple routes of facility networks crisscrossing one another.

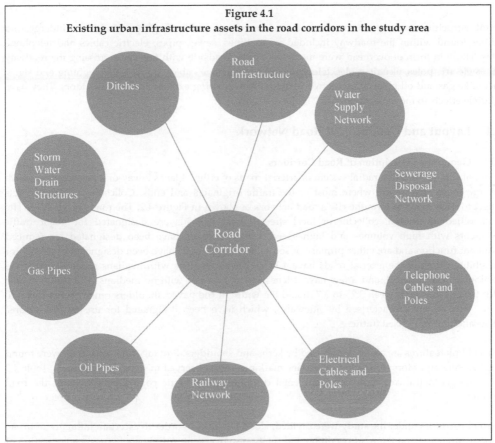

Figure 4.1
Existing urban infrastructure assets in the road corridors in the study area

Source: Field data survey (2007).

The inventory survey showed the type of urban infrastructure in each road corridor as summarised in Table 4.1.

Table 4.1
Urban infrastructure assets along the road corridors

S/N	Corridor Name	Water Supply	Sewer Network	Electric Supply	Telephone Network	Gas Pipe	Railway	Oil Pipe
1	Bagamoyo Road	√	√	√	√	X	X	X
2	Morogoro Road	√	√	√	√	X	X	X
3	Nyerere Road	√	√	√	√	X	√	X
4	Kilwa Road	√	√	√	√	√	√	√
5	Mandela Road	√	√	√	√	X	√	X

Key: √ - exists, X- not exist
Source: Field data survey (2007)

The infrastructures were found to exist in form of overhead and underground facilities. Underground facilities found within the roadway included water pipes, sewer pipes, electric cables and telephone cables. Those in form of overhead were mostly along the roadside with few cases crossing the roadway such as electric poles, electric cables, telephone poles, telephone cables, traffic lights, lighting and signs. Others like gas and oil pipes, fibres network and railway crossing exist in few road corridors. They have negligible effects to movement and safety in the road corridor.

4.1.1 Layout and Condition of Road Network

(i)   Geometric Description of Road Corridors

Dar es Salaam consists of a radial system of arterial roads of either 4 lanes 2 way or 2 lanes 2 way, which converge in the city centre where most of the traffic originates and ends. Collector and local streets connect to such roads to form the city's road network as shown in Figure 4.2. The roads in the study site are classified as arterial, collector or local streets. Arterial roads are designated for major traffic movements with high volumes and high speed. Collector roads have been designated for reduced movement function and are either primary or secondary. Local roads have been designated primarily for accessibility. Most of the arterial roads have 4 lanes 2 way facilities with medians separating the two directions of traffic travel, and a few have 2 lanes 2 way facilities with no medians. The common lane width for arterials is between 3.25 to 3.7 m and the width of the paved shoulders ranges between 1.0 to 2.0 m. Arterials are characterised by sidewalks, which have been designated for use by pedestrians, cyclists and non-motorised traffic.

The road link features are also characterised by kerbs and shoulders. Frontage and separators were found along Nyerere and Morogoro roads, while lane markings were observed in some road sections. Table 4.2 shows the geometric conditions, environmental conditions and traffic control conditions of the road corridors.

The road corridors under the study were divided, along the links in order to ascertain influence of urban infrastructure assets on the road condition and safety of movement. A link is a homogeneous section of the road excluding the main junction, which has uniform cross-sectional characteristics such as number of lanes, road width, median, and shoulders as well as uniform adjacent land use. It contains less important minor junction and accesses. A junction (Node) is the section of the road where at least one major road intersects with one or more main or busy minor roads (carrying AADT higher than 500) and extends approximately up to 30 m from the stop line, depending on assumed influence on the junction layout.

A total of 49 link study sites were selected on the city arterial roads, which represent various geometric and traffic conditions in the city. They included 19 undivided road sections and 30 divided road sections. The length of link sections varies between 0.40 km and 3.21 km with an average of 0.92 km. In summation, the total length of road sites amounted to 53.8 km with clear demarcations of the begining and end of the selected roads under the study. Table 4.3 shows the number of link sites and type of carriageway found in the road corridors.

Figure 4.2
Typical links and junctions in Dar es Salaam

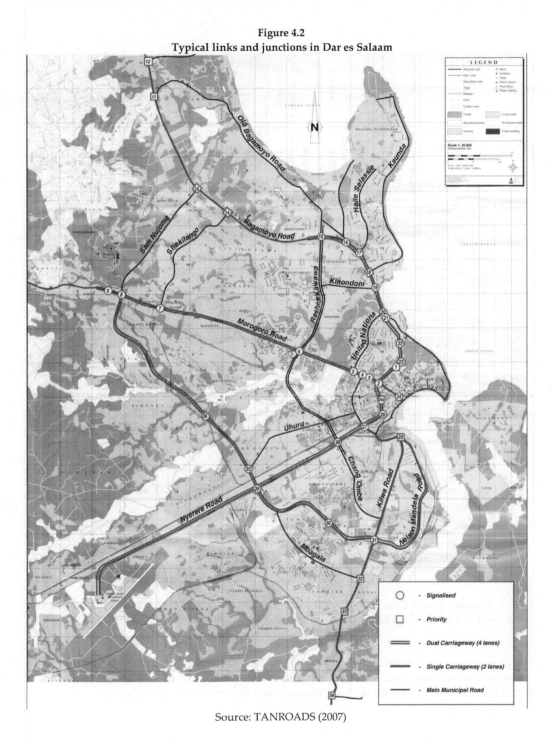

Source: TANROADS (2007)

Table 4.2
Study sites characteristics

STUDY SITES			NSSF	Ubungo	Tanganyika Motors	Morocco	Mwenge	Chang'ombe	TAZARA	Uhasibu
Road/Location			Morogoro, 0.9 km, West of CBD1	Morogoro, 9.1 km, West of CBD1	Bagamoyo,1.7 km, North of CBD1	Bagamoyo, 6.6 km, North of CBD1	Bagamoyo, 10.9 km, North of CBD1	Mandela, 9.1 km, West South of CBD1	Nyerere, 10 km, West South of CBD1	Kilwa, 6.5 km, South of CBD1
Geometric conditions	Cross sections	Road type	2/2UD²	4/2 D³	2/2UD²	4/2 D³	2/2UD²	4/2 D³	4/2 D³	2/2UD²
		Carriage way width	6.75m/ Two ways	7.4m/ Two ways	6.4m/ Two ways	7m/ Two ways	7.8m/ Two ways	7.8m/ Two ways	7.8m/ Two ways	6.4m/ Two ways
		Outer shoulder width	2.45m	1.60m	1.65m	0.5m	1.60km	1.70km	2.2km	2.0km
		Median width	-	3.0m	-	1.5m	2.0m	3.0m	1.5m	-
		Walk way	-	3.0m	-	3.0m	-	-	3.5km	-
	Alignment/ Terrain		FS	FS	FS	FS	FS	FS	FS	FS
	Sight distance		Infinity	Infinity	Infinity	Infinity	Infinity	Infinity	Infinity	Infinity
Environmental conditions	Off road facilities		R	Street Vendors	C	R	R	I	I	R/C
	Weather conditions		D/DP	D/DP	D/DP	D/DP	D/DP	D/DP	D/DP	D/DP
	Urban / Suburban		Urban	Urban	Urban	Urban	Urban	Urban	Urban	Urban
	Pavement conditions		PS	PS	PS	PS	PS	PS	PS	PS
Traffic control	Speed Limits		50km/h	30km/h	50km/h	Not posted	Not posted	Not posted	50km/h	50km/h
	Pavement markings		Not Marked	Marked	Not Marked	Marked	Marked	Not Marked	Marked	Not Marked

NOTE:

CBD1 = Central Business District, UD2 = Un-Divided roadway, D3 = Divided roadway, R=Residential, I=Industrial, C=Commercial, D/DP = Day/Dry Pavement, PS = Paved and Smooth, FS=Flat and Straight.
Source: Field data Survey (2007).

Due to inadequate information on the geometry and facilities of road sites from the road authority, actual measurements were taken and visual inspection was carried out to collect the necessary infrastructure data along each road corridor site in Dar es Salaam. Data collection on link sites included:

- Carriageway type (divided or undivided)
- Road surface condition
- Length of link
- Lane width
- Number of lanes
- Median width (if the road is divided)
- Width of sidewalk and surfing
- Shoulder condition
- Presence or absence of raised curb at the edge of the road
- Number of minor junctions per km (where AADT coming from the minor road is not more than about 500)
- Horizontal alignment
- Vertical alignment
- Type of separation
- Underground infrastructure
- Roadside Infrastructure and
- Manholes condition on the road surface.

Link sites were coded or given reference numbers to reflect their proper location.

Table 4.3
Summary of inventory survey

Road Features	Bagamoyo Road	Kilwa Road	Mandela Road	Morogoro Road	Nyerere Road
Pavement length (km)	37.46	16.88	20.78	32.84	12.52
Pavement area (m2)	290,509	111,352	400,385	492,563	266,409
Shoulder length (road km)	37.46	16.23	20.78	30.53	12.52
Shoulder area (m2)	282,982	86,356	189,460	292,994	125,190
Number of Culverts	7	27	6	49	11
Number of Sub-links with Side	47	21	24	41	16
Number of Metre Drains	0	5	0	18	0
Number of Bridges	1	3	2	4	0
Number of Signs	102	31	23	110	53
Guard Rail length (m)	2,735	0	1,439	7,028	4,667
Number of Junctions	6	68	17	44	2
Number of Rail Crossings	0	2	4	0	2
Number of Sub-links	47	21	24	42	16

Source: Field data Survey (2007)

(ii)   Road Condition

Road condition was determined by assessing elements of roadway, which included road surface, shoulder, walkway and side slope. Road surface was characterised with distresses such as ruts, potholes, cracks and patches. Also surface roughness was measured.

Shoulder condition was determined by assessing the extent of presence of edge brake and potholes. Assessment of walkways was done by observing a few potholes, edge break, settlement of slabs and encroachment of motorised traffic. Furthermore, for the side slope, assessment was carried out for erosion especially on unpaved roads.

The combined extent of failure was used to rate the roadway into good, fair or poor as classified in the Road Maintenance Management System (RMMS). Results are as shown in Figure 4.3 for the studied road corridors. Parameters for rating road condition are given in chapter 7.

Figure 4.3
Road corridor condition in studied area

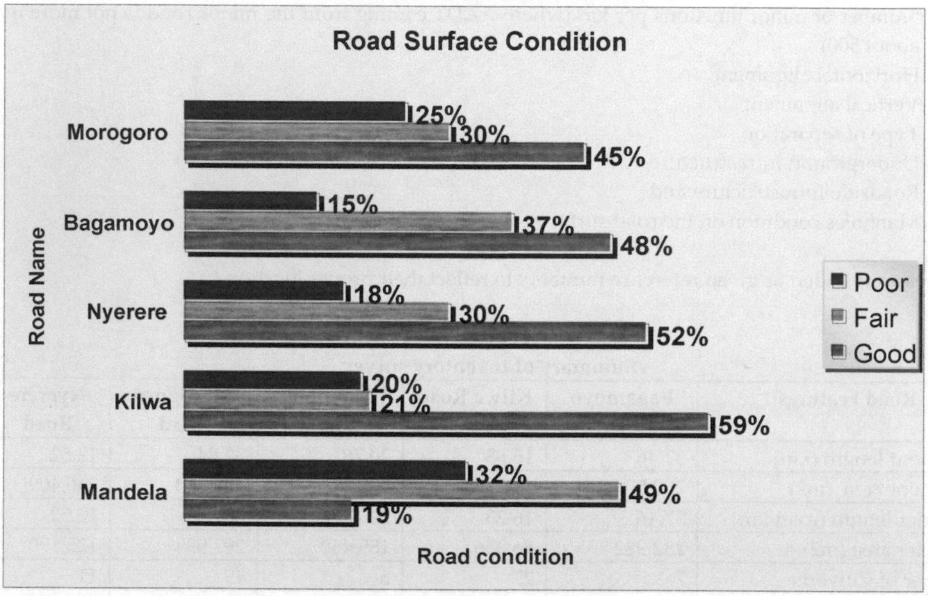

Source: Field data survey (2007)

Figure 4.3 shows three types of road corridor conditions established after assessing conditions of the five selected roads. The road was assessed to be good when there were no visible defects, but it was assessed to be fair if it had low frequency of defects with medium severity. The road condition was said to be poor when it had medium frequency of defects with high severity or when it had high frequency of defects with medium severity. The good road condition varied from 45 to 59 percent, fair condition varied from 21 to 49 percent and poor road condition varied from 15 to 32 percent. However, further link-wise assessment was carried out on how the rated road condition within a particular section of the road was associated to the condition of other infrastructures within the road corridor.

4.1.2   Layout and Condition of Storm Water Drains

The layout of storm water drains system in Dar es Salaam was found to form an integral part of elements existing along the road layout. They comprised roadside drains and storm drains from the premises adjacent to the roads. Road side drains not only served as drains for the road surfaces but also as collectors and disposal for discharges from drains originating from the premises.

Storm water drains exist as open channels, covered channels or buried pipes that are mostly laid parallel to the carriageway adjacent to road shoulders and are located either within the road shoulders or within the carriageway. The buried and covered routes comprised also of manholes that are either used for trapping or as joints. Elements that were assessed included soundness of lining, presence of infrastructure facilities that were laid along and across road corridors plus silt.

The survey revealed that storm drains from the adjacent premises were randomly connected to roadside drains and in most cases the connections were so poorly finished that they became the cause of weakening linings of roadside drain. Furthermore, they were connected at wrong slopes and also to the low capacity channels, which resulted interfered with steady flow and caused stagnation as well as frequent over flooding. When interviewed, city council officials said that lack of coordination among road designers and land planners was the source of poor drainage.

Findings from interviews revealed that there was no robust monitoring and control measures instituted preventing land users from randomly connecting to the side drains. Also unplanned land use especially development of buildings within open spaces and basins blocked natural water courses and thus escalated the effects. Furthermore, there was very little adoption of the low impact development approach among developers that could help to relieve constrains of disposal for most storm water originating from the developed premises through infiltration within the plots.

The road authority revealed that Dar es Salaam experienced serious problems of drainage. During rains, parts of the city got flooded to the point of impairing efficiency of certain production activities in the city. For example impacts of a malfunctioning drainage system are shown in Figure 4.4, which was one of the typical cases that affected the roadway. Frequent clogging of road storm water drains was associated with the infrastructure laid along and across the drains, which blocked the drains.

Figure 4.4
Drainage problems along Mandela Road, Dar es Salaam

Source: Field Data Survey (2007).

It was reported that water overflows on road corridors due to leakage of underground pipes and due to poor restoration of cut road sections for underground infrastructure operations. Although most drains in the CBD are covered, drains are silted, so they fail to drain water from the road surfaces. Part of thereason was that such drains were used as infrastructure passages. In some places, mainly outside the CBD, there are open drains and debris obstructs the drainage system. This, plus uncoordinated operations, planning and poor design of the urban infrastructure contributed to the observed problems regarding the movement and safety of people and vehicles.

Figure 4.5
Effect of water on the road surface

Source: Field data survey (2007).

Figure 4.5 shows effects of water on the road surface and rapid increase in the degree of rutting which often leads to cracking and breaking up of the pavement. If excessive, it can reduce serviceability and vehicle travel speed. Very severe cases may result to accidents and high vehicle operating costs.

4.1.3   Water Supply Network

Observation showed that water pipes are placed under or above the road corridors. When interviewed, DAWASA officials said that some main water pipes were placed early in 1970s; therefore, they are old. Planning for refurbishing and improving the network is done slowly not at the same pace as the increase in demand and change of technology.

Figure 4.6
Condition of underground water pipes along Bagamoyo Road

Source: Field survey data (2007)

The information obtained from DAWASCO officials further revealed that there was no system set to track leakage and burst of the water pipes; instead customers or passers-by reported problems to DAWASCO authority. There is no regular maintenance to the water supply network, only repair is done on emergencies. The capacity supplied per day is 281,000 cubic litres; however, only 127,000 litres reach consumers, while 45 percent of water is lost through illegal diversions and leakage. The three water sources available in Dar es Salaam are Kizinga, Upper and Lower Ruvu River intakes (DAWASCO Annual Report, 2007).

Figure 4.6 shows underground water pipe networks concentrated under and along the road corridors. Different water pipe sizes were seen under the road corridors in the study area. The system of underground water pipes along the road corridors in Dar es Salaam is illustrated in Figure 4.7.

Figure 4.7
Underground water pipes along road corridors in Dar es Salaam

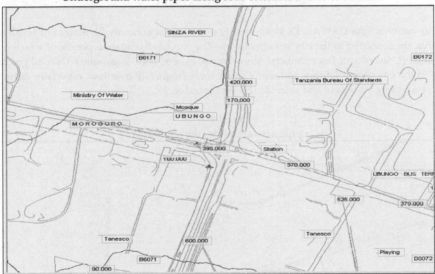

Source: DAWASA (2007)

4.1.4 Sewer Management Network

It was learnt from DAWASA authorities that the sewer system had been established between 1940 and 1960. It implies that the water supply network was fixed before independence. The sewer pipes are placed in the middle of the carriageways, the old system joints are at 90 degrees when connecting to customers and the manholes are placed in the middle of the road surface. It becomes difficult and hazarduos when attending emergency cases of sewer network leakages and blockages.

Figure 4.8
Sewer pipe leakages at Magomeni and Red cross along Bagamoyo and Morogoro Roads

Source: Field survey data (2007)

According to statistics from DAWASCO, the authority managing sewer, only 21 percent of residents in the study area are connected to the city sewer network. The city has 8 oxidation ponds, of which only 4 are in operation (University of Dar es Salaam, Kurasini, Mikocheni, and Vingunguti). Over 80 percent of households in the city use pit latrines and septic tanks; these frequently overflow, especially during the rains, thereby affecting movement and safety in the road corridor.

Figure 4.9
Sewer pipe network under the study area

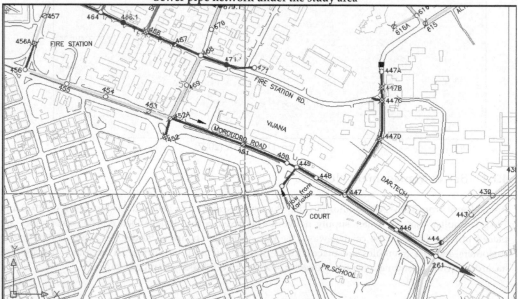

Source: DAWASCO (2007)

Also, like in the water supply network, the underground sewer infrastructure is outdated and dilapidated causing frequent overflow of waste onto the road surface, thereby affecting movement and safety as seen in Figure 4.8. There is no regular maintenance done but rather what is done is ad hoc repair, where by a system is repaired when a leakage or blockage in the network has been reported.

It was also reported during the interviews that DAWASCO has insufficient knowledge about the underground sewer pipe network. Currently, there is no tracking system to locate the sewer pipe leakages and bursts. Sometimes they only come to know about the existence of the pipes and the problem after a leakage or burst has been reported. Figure 4.9 shows the sewer network in Dar es Salaam.

Table 4.4 provides an inventory of sewer manholes placed along Morogoro road. Since many of the sewer manholes are placed in the middle of the carriageway, observation from surveillance cameras showed trucks parked in the middle of the carriageway empting the blocked sewers and prolonged leakages on the road surface.

Table 4.4
Manholes on the road surface along Morogoro Road

Pipe No.	1	2	3	4	5	6	7	8
Start Manhole	K454.1	K453.1-1	K453.1	K452.1-1	K452.1	K451	K451.1-1	K450.1
End Manhole	K453.1-1	K453.1	K452.1-1	K452.1	K451	K451.1-1	K450.1	K449.1
Diameter (mm)	300	300	300	300	550	550	550	550
Material	PVC	PVC	PVC	PVC	GRP	GRP	GRP	GRP
Length (m)	33.4	58	46	63	44.2	54.1	60.4	54.1
Depth (Start)	4.8	4.803	4.807	4.71	4.637	4.441	4.3	4.137
Function	Sewer Trunk	Sewer Trunk	Sewer Trunk	Sewer Trunk	Sewer Trunk	Sewer Trunk	Sewer Trunk	Sewer Trunk
Start Ground Level (m)	95.320	95.236	95.090	94.870	94.570	94.180	93.955	93.680
Start Invert Level (m)	90.520	90.433	90.283	90.160	89.933	89.739	89.650	89.543
Slope1 in	370	370	250	250	440	440	500	500
Bedding Type	Sand bedding	Sand bedding	Sand bedding	Sand bedding	Sand bedding	Sand bedding	Sand bedding	Sand bedding

Source: Field data survey (2007)

4.1.5   Electric Power Supply Network

Electric power supply network in the road corridors is found underground, on the surface and above the surface. Located under the ground are electric cables, on the road surfaces are manholes and the control point and on the roadside are electric cables and poles.

The Tanzania Electric Supply Company (TANESCO) is a utility company owned by the Government of Tanzania. The core business of its activities are generating, transmitting, distributing and selling electricity. It also manages the most dependable power stations, namely, Mtera, Kidatu, Pangani, and Nyumba ya Mungu. TANESCO produces 347 megawatts (MW) of electricity against a country maximum demand of 550 MW.

Commercial cities like Dar es Salaam, Mwanza and Arusha accommodate large-scale consumers of electricity. Therefore the pattern increases interaction of underground and roadside infrastructure in the road corridors. The electric cables network in the road corridors are extended to traffic light and street light poles.

Figure 4.10
Relocation of electricity poles and cables

Source: Field data Survey (2007).

An example of relocation of electric poles and cables as shown in Figure 4.10 in the study area is done to allow rehabilitation works of a road section. But since the network is not well known to the operator, un-relocated cables are cut by construction equipment.

4.1.6   Telecommunications Network

Telecommunications network is managed by the Tanzania Telecommunication Company Limited (TTCL) without outsourcing operational services. Main operational services are provision of connectivity (building of access lines) from the distribution points within the core network to customers' premises, attending network faults, and undertaking preventive maintenance.

The network is comprised of cable network and wireless network. Of interest to this study was the cable network, which physically occupies space and its location and layout interact with other urban infrastructure networks. The cables are either suspended between poles or buried underground. The network is also comprised of joint and distribution boxes placed or protruding above ground level for jointing or branching of cables.

Cables have been laid or suspended along the shoulders or in other areas crossing the road network. Others have been placed far from the shoulders at variable offsets from the centre of the road, depending on availability of space limited by constraints caused by existence of other infrastructure sharing the same passages.

Telecommunication cables were found to be either of fibre optic or copper wire. The fibre cable is a bundle of thin glass like fibre in which data travels at the speed of light and has very high capacity. Cable sizes are defined according to capacity of connecting to customers rather than diameter size. Average cable diameter is 50 mm. Cables are laid in the PVC ducts of 100 mm diameter, while overhead cables are of 40mm diameter.

Underground cables are laid either as direct inburied in the ground or passed within the underground laid PVC ducts at a depth from 0.9 to 1.2 m (as shown in Figure 4.11). Overhead cables are suspended along treated wooden poles at heights of 6 m to 8 m, erected along the shoulder or crossing the road. The interval between the poles is a minimum span of 50 m. But wooden poles are not erected all along the whole length of the roads, they are erected only where there are customers to be served from nearby distribution points.

Table 4.5
Telecommunication cables and poles network on road sections

Road Name	Road Section	Underground Cables		Overhead Cables		Poles No.	
		EPS	TN	EPS	TN	EPS	TN
Bagamoyo	Ohio	√	√	√	√	169	66
	Ali Hasani Mwinyi	√	√	√	√		
	New Bagamoyo	√	√	√	√		
Morogoro	DSM-Fire	√	√	x	X	120	172
	Fire-Magomeni	√	√	x	X		
	Magomeni-Ubungo	√	√	√	√		
Nyerere	Ohio-Uhuru	√	√	x	X	377	273
	Uhuru -Changombe	√	√	x	X		
	Changombe-Tazara	√	√	x	X		
Kilwa	Kivukoni-Railway	√	√	x	X	115	72
	Railway-Bandari	√	√	x	X		
	Bandari-Uhasibu	√	√	√	√		
Mandela	Mwenge-Ubungo	√	√	√	√	147	98
	Ubungo-Tazara	√	√	√	√		
	Tazara-Bandari	√	√	√	√		

Source: Field data survey (2007).

Table 4.5 shows the distrubution of underground and overhead telephone cables in the identified road sections.

Figure 4.11
Operations of telephone cables in the road corridors

Source: Field data survey (2007).

4.1.7   Assessment of Relative Location of Infrastructure within Road Corridors

To understand spatial interactivity of existing infrastructure, location of network facilities were considered relative to common reference points to reveal potential of conflicts associated with their layout. Magnitude of conflict was rated by closeness to each other, that is, proximity. Proximity was determined based on the limit of distance within which effects resulted from excavation can be propagated and affect substantial conditions of other peripheral infrastructures. Such distance is based on what was revealed by Symons *et al.* (1982) that zones of movement normal to the trench extended to a distance of 2–2.5 times the trench depth on each side, with lateral movements extending slightly further than settlements. Thus, for a water pipe placed at the depth of 1.5 m, facilities that are located within a distance of 2 to 2.5 times this depth are within proximity.

Table 4.6
Urban infrastructure proximity rating for Morogoro Road

	Link	Sewer	Water pipes	Telecoms	Electricity	Proximity rating
Off set in metres from the edge of the road centre line	1	0.0	1.0	1.5	2.0	1
	2	0.0	1.0	2.0	2.5	1
	3	1.5	2.0	3.0	3.0	1
	4	1.5	2.0	3.0	3.0	1
	5	0.0	1.0	2.0	2.0	0
	6	0.0	0.0	1.0	2.0	1
	7	1.0	1.0	3.0	2.0	0
	8	1.5	2.0	3.0	3.0	1
	9	1.5	3.0	4.0	4.5	1
	10	3.0	3.0	4.0	5.0	0

Source: Field data survey (2007)

Relative location of the main infrastructure assets measured from the common reference line within the road corridor is summarised in Table 4.6 for Morogoro road. The indicated sections were chosen to group locations that had homogeneous corridor width.

The distance between adjacent infrastructure routes was determined and categorised in respect to proximity, based on limit of influential distance given by depth. In order to rate the extent of closeness, the existence of proximity was assigned an arbitrary number 1 and non-existence was assigned 0.

Data from Table 4.6 show that relative location of infrastructures from the reference points within the roadway kept on changing, regardless of their location with constant corridor width. Also the relative location of infrastructure network layout was not constant. They kept on crisscrossing like a web and they were essentially for underground facilities. It was established that the infrastructure facilities were placed inconsistently along the road corridor. They kept on interchanging location horizontally and randomly, crossing each other within very close clearance depth. Therefore, it was established that there was inconsistency in layout of the facilities.

The reason for such inconsistency was revealed through interviews conducted at TANROADS and DAWASA. It was revealed to be significantly due to the procedure used to allocate space for underground infrastructure installation. Procedures used to locate the infrastructure network in the road corridors were first come, first served. Installers of facilities acted at liberty in changing the layout at their own convenience without abiding by uniformity in layout, because there was no functioning control mechanism. As a result the installed infrastructure facilities were not consistently located within the road corridor, so they overlaped each other. Hence, increased conflict and failure of infrastructure when one operator had to work on another as infrastructure network.

It was revealed from infrastructure layout data that infrastructure networks within the road corridor are close distance apart. The observed infrastructure facilities were found to be laid in proximity to each other to the extent of being miss-identified especially with lack of locator aids even within the area with wider corridors. The maximum and minimum distances that were observed among the infrastructure facilities within the study corridors were 7 metres and 1 metre respectively. Such findings imply that apart from the urban infrastructure facilities sharing the corridor, they were also geographically related in terms of proximity in layout of their routes.

Further assessment of relative layout showed digit 1 being the score with maximum frequency (mode) obtained from rating of closeness of infrastructures along the road corridors. It indicated that there was significant statistical evidence not to reject the construct that urban infrastructures exist in proximity.

4.1.8   Operational and Faults Interactivity among Infrastructures in Dar es Salaam

Operational and fault of dependence in urban infrastructure within the road corridors of the study area were assessed to establish the nature of interaction among infrastructures associated with normal operation activities also emanating from occurrence of faults in one type of infrastructure. The assessment also was to find out whether the interactivities resulted into positive or negative impacts and the likely source of such nature of interactions.

Operational activities of urban infrastructure assets were mainly carried out by owners themselves, except a few specialised activities that were outsourced. The outsourced activities required adequate resources (human resources and equipment), which were not readily available within the parent firm.

Existence of operational interaction among the urban infrastructure operational processes or duties that were conducted within the corridor in a way of optimising utilisation of utility capacity was assessed. The assessment entailed determining whether or not there was any process accomplishment constraints caused by operational process undertaken within the other infrastructures. Operational duties that were assessed included services connectivity to consumers, network monitoring, data management, routine maintenance and fault repairs.

All such activities were assessed on their significant interaction with operations in other infrastructure facilities through space displacement. Undertaking of such activities within the shared locations resulted into reduction of service delivery through constriction of the shared spaces. A typical case was that of routine maintenance of the sewerage system through manholes within the carriageway, which resulted into reduction of carriageway capacity by constriction or closure of one lane. The more the number of manholes attended concurrently within the carriageway, the higher was the interaction.

Findings from network layout from the firms' database showed that firms for urban infrastructure assets randomly shared location in the road corridors and in some parts, they were not easily identifiable. Such findings support what was observed in the road corridor inventory survey. In terms of effectiveness, it was noted that, it had to take long time to identify faults and effect restoration of services, especially for sewer and water supply.

Operational activities, especially maintenance and repair works were prone to damage other infrastructure networks due to improper or lack of universal documentation of all network layout along the road corridors. Tracing of faults by trial and error within the corridor occupied substantial space, took a long time, and ultimately affected traffic and safety of movement.

In the assessment of faults interaction among infrastructures, a series of faults occurring in one infrastructure were related to those recorded in others to establish existence of physical and logical dependence. Elements that create functional dependences like existence of supplier-consumer linkage and logical control were also analysed.

4.1.9   Interactivities in Urban Infrastructure Operations

Operational interactions among the infrastructures were conceptualised as shown in Figure 4.12. Due to the potential of the observed proximity, activities related to one infrastructure were recorded to impact other infrastructures within the road corridor environment. Interaction in terms of cause and effect to each other were identified and recorded as shown in Table 4.7 for Morogoro road. Records were from 2005 to 2009.

Water supply operational activities and leakages due to faults were recorded as adversely affecting other infrastructures in the road corridors. The problem was highly escalated by lack of reliable pipe networks. Effects were critical when there was inevitable relocation of other utilities to allow for rehabilitation or upgrade of the road network. Sometimes, other urban infrastructure networks were cut by construction machinery.

Figure 4.12
Interactivity of infrastructure on each other

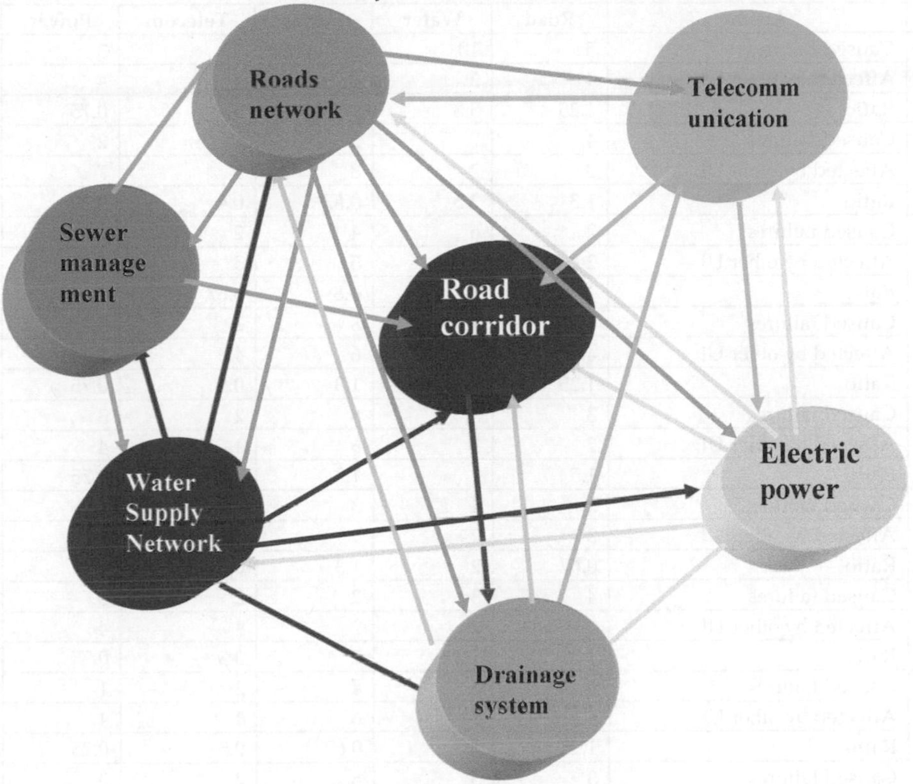

Source: Field data survey (2007)

Geographical interdependence existed as the consumers of services share locality. The Need for consumption of all services by the same user necessitates physical location of networks of utilities either in proximity or within the premise of the consumer so as to avail the required services. Hence, inevitably converged and were confined within the same limited space.

Furthermore, when utilities provide services to other utilities within their location (supplier and consumer relationship); it is necessary that such utilities get located in the same area to form interdependent network. For example, an electric service provider provides energy to street lights, traffic lights and sign boards. Also the water supplier avails services through a network of water pipes to fire hydrants located along the road shoulders, which are accessible by vehicles.

Table 4.7
Number of times infrastructure leading to failures

Link	Action	Road	Water	Sewer	Telecom	Power
1	Caused failures	5	10	8	2	3
	Affected by other UI	4	3	6	4	4
	Ratio	1.25	3.3	1.3	0.5	0.75
2	Caused failures	4	7	2	2	2
	Affected by other UI	3	2	3	5	7
	Ratio	1.3	3.5	0.67	0.4	0.29
3	Caused failures	2	6	4	2	1
	Affected by other UI	2	1	5	3	1
	Ratio	1	6	o.8	0.67	1
4	Caused failures	5	10	8	2	3
	Affected by other UI	4	3	6	4	4
	Ratio	1.25	3.3	1.3	0.5	0.75
5	Caused failures	3	3	8	2	3
	Affected by other UI	2	3	6	4	4
	Ratio	1.5	1	1.3	0.5	0.75
6	Caused failures	2	4	3	1	0
	Affected by other UI	3	2	2	2	1
	Ratio	0.67	2	1.3	0.5	0
7	Caused failures	4	3	2	2	3
	Affected by other UI	3	2	2	4	4
	Ratio	1.3	1.5	1	0.5	0.75
8	Caused failures	5	6	4	2	1
	Affected by other UI	4	3	6	4	4
	Ratio	1.25	2	0.67	0.5	0.25
9	Caused failures	3	3	8	2	3
	Affected by other UI	2	3	6	4	4
	Ratio	1.5	1	1.3	0.5	0.75
10	Caused failures	2	4	3	1	2
	Affected by other UI	1	2	2	2	1
	Ratio	2	2	1.5	0.5	2

Source: Field data survey (2007)

(i)   Effect of urban infrastructure operation on the road network

Figure 4.13 shows a road section cut in the middle of the carriageway for an underground sewer pipe repair at Ohio Street along Bagamoyo Road. An interview with contractors dealing with such repairs showed that poor repair techniques, poor workmanship during excavation, lack of prior and proper signing on work zone to the public led to damaging water pipes and telecommunication cables present in the work zone. It also affected services delivery and safety of movement in the road corridors. Excavation and reinstated road sections also affected the traffic movement and queuing on the road corridors. It was found out from the interview, inventory survey and surveillance cameras that the urban infrastructure operation in the road corridor damaged of the road surface and road network structure.

Figure 4.13
Road section cut for underground sewer along Bagamoyo Road

Source: Field data survey (2007)

For example in year 2006, a major repair of underground sewer pipes was implemented along Ohio road through cutting of the newly placed surfacing. However, due to poor workmanship during reinstating, the section failed and caused ruts and potholes on the road surface. The first failure was observed in year 2007 and was repaired. Another failure occurred in year 2008.

Figure 4.14
Road surface failures along Bagamoyo Road

Source: Field data survey (2007)

Dar es Salaam City Council provided records of affected parts of the road network and DAWASA provided a number of damaged underground sewer pipes in the middle of carriageways. Figure 4.14 shows road surface failures of the road section along Bagamoyo road due to underground infrastructure repair. Cutting of road layers (weakening road layers and damages infrastructure) coupled with poor workmanship, led to failure of road surface as observed in all sections cut for underground infrastructure repair.

Other reported operational conflicts of the urban infrastructure included the following:
- Impairing level of services of urban infrastructure due to frequent damages and prolonged restoration time of the failed facilities;
- Induction of propagated failure due to poor repairing techniques;
- Lack of database for existing infrastructure facilities in the road corridor;
- Lack of coordination among infrastructure operators; and
- Lack of financial resources.

4.2   Interdependencies in Infrastructure Development and Management

Apart from operational approaches, the study sought to find out whether or not development and management approaches of urban infrastructures had significant influence on observed conflicting interdependence. As pointed out by Ferreira and Flintsch (2004), the public and private agencies that own urban infrastructures recognise the challenges posed by interaction and focus on system approaches into management of assets with a milestone of achieving Integrated Development and Management of infrastructures. An integrated infrastructure development and management system is a systematic approach that achieves coordinated planning and programming of investments or expenditures, design, construction, maintenance, rehabilitation, renovation, operation, and in-service evaluation of physical facilities. The system includes methods, procedures, database, policies, and decision means necessary for providing and maintaining infrastructure at a level of service acceptable to the public or owners.

The main indicators that were assessed included presence of elements of total corridor management, consistence in policies governing the agencies and presence of urban infrastructure database. Furthermore, information concerning drivers for formulation of development and maintenance plans, presence of effective coordination and strategic plans were also assessed. Information was collected through interviews, questionnaires and some input from documentary review.

4.2.1   Development Approaches

Urban infrastructure system development can be clearly explained by looking into salient features of main stages of their development, namely; planning, designing and implementation. Assessment of planning undertaken by most firms was seen to be ineffective due to resource constraints and such plans had not been drawn from feedback obtained from previous strategies due to lack of effective monitoring and evaluation. For example, the management in water supply admitted that they did not have long term plans that could be used as sources for drawing short-term plans. They also lacked maintenance plans because they were incapable of monitoring and evaluating performance of delivered services.

Stakeholders were of the opinion that provision of urban infrastructure was driven by unpredictable rises of consumer demand in an ad hoc way rather than being in line within strategic plans. Also others were politically driven, predominantly in government agencies. By a cross-examining some agencies revealed that planning was not coordinated among urban infrastructure stakeholders (developers and consumers). This resulted into wastage of resources that would otherwise supplement one another (like provision of universal underneath ducts to allow for underground cable and pipe lines to cross roads without requiring trenching).

There was also a tendency of carrying out ad-hoc planning to cope with inevitable infrastructure network requirements driven by others, for example relocation of utilities caused by developments carried out by other operators sharing the same road corridor, especially on road up grade.

Another obstacle was limited source of funds, leading to unreliable financial planning, insufficient provisions and a track log of unfinished projects. Consequently, synchronisation of implementation among the operators was rendered impossible.

In designing an urban infrastructure system, there was a problem caused by differences in standards adopted among service providers. Furthermore, the assessment revealed that design and respective implementation became immediately ineffective due to dynamic land usage and spontaneous uncontrolled growth in urbanisation. Ultimately, redesigning was frequently needed due to limited expertise.

4.2.2   Infrastructure Management

There are at least seven Ministries involved through their agencies and their interdependence cuts across each other. They are the Ministry of Infrastructure Development (MOID), Ministry of Communications and Transport (MOCT), Ministry of Home Affairs (MOHA), Ministry of Finance and Economic Affairs (MOFEA), Prime Minister's Office, Regional Authority and Local Government (PMORALG), Ministry of Energy and Natural Resources (MOE) and Ministry of Water and Irrigation Services (MWI). In addition, there are private companies such as communication services providers.

One of the investigated aspects was existence of conflicts within the corridor as a result of decentralisation. In order to improve performance the role of operating and managing the infrastructures was decentralised to lower levels through agencies and authorities, by delegating some functions. Decentralisation is being considered to be an appropriate approach into achieving effectiveness. It creates more flexibility and it is quick in responding to changing needs of customers (Osborne and Gaebler, 1992).

There are conflicting authorities in utilisation of corridors among the government agencies established under Parliament Acts, these include, Tanzania Telecommunication Company Limited (TTCL), Dar es Salaam Water and Sewerage Authority (DAWASA), Tanzania National Roads Agency (TANROADS) and Municipal Council. While TANROADS is responsible for managing road corridors, its control is limited by power vested to other public firms through different Acts.   The firm's managing urban infrastructure in the road corridor are Tanzania National Roads Agency (TANROADS), Dar es Salaam City Council (DCC), Dar es Salaam Water and Sewerage Authority (DAWASA), Dar es Salaam Water and Sewerage Corporation (DAWASCO), TANESCO, Telecommunication Companies, Tanzania Railways Corporation (TRC) and Tanzania Petroleum Development Corporation (TPDC). Figure 4.15 lists various operators who, in one way or another, manage urban infrastructure.

Figure 4.15
Various operators managing urban infrastructure

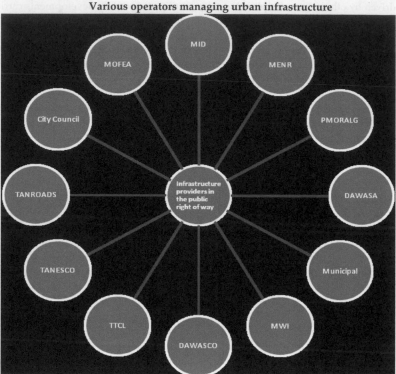

Source: Field data survey (2007)

(i)   Management of the Road Corridor

Road corridors in Dar es Salaam City are under the management of the Ministry of Infrastructure Development (MOID), Prime Ministers' Office, Regional Authority and Local Government (PMORALG). Categories of the road which are under MOID in the city are all arterial roads and the rest are under PMORALG.

Tanzania National Roads Agency (TANROADS), which is an Executive Agency of the Ministry of Infrastructure Development, is responsible for daily management of the country's trunk and regional road network where arterial roads in the city lie. Its primary functions include management of maintenance and development works including operations on the network. The networks under PMORALG are managed by City councils and Municipal councils.

The study revealed that although TANROADS is vested with authority to manage the corridors, it lacks an effective structure to undertake control and monitoring of usage of the corridors. During the interviews the road authority pointed out the following facts about TANROADS.

- They own the corridors, but lack knowledge of the detail of all utilities, routes and facilities laid in the corridor;

- The authority has not yet considered the need of having a database of types of infrastructure existing within the corridor; and
- Coordination duties for corridor utilisation have not been streamlined.

(ii)   Management of Water Supply and Sewer Network

Management of water supply in Dar es Salaam city as already mentioned is done by Dar es Salaam Water and water supply and waste disposal services. They are working under the umbrella of Dar es Salaam Water and Sewerage Corporation (DAWASCO), the sole provider of water supply and sewerage services in Dar es Salaam city and parts of Coast region. The Corporation is responsible for management, operation and maintenance of Sewerage Authority (DAWASA). Both DAWASA and DAWASCO are Executive Agencies under the Ministry of Water (MW).

(iii)   Management of Electric Power Supply Network

Tanzania Electric Supply Company (TANESCO) is a utility company owned by the Government of Tanzania. Being a sole provider, its core business activities are generating, transmitting, distributing and selling electricity. It also manages the most dependable power stations. TANESCO is under the Ministry of Energy and Minerals (MEM).

(iv)   Management of Telecommunication Networks

Four firms were found to own telecommunication cable routes within the road corridors. Each firm processes consent and lays cables independent of the other firms. These are installed in interactive corridors. Interviews revealed that though they were required to get prior consent from TANROADS and work in liaison with other utility providers, in practice this did not happen because TANROADS does not have regulatory resources.

Lack of coordination between the authorities managing the urban infrastructure network, has led to the following shortcomings:

- Poor planning, design, maintenance and operation of the urban infrastructure, since they were implemented differently;
- Relocating urban infrastructure to allow rehabilitation of road sections, which is very costly;
- Delays in project implementation due to lack of communication among operators;
- Reduction in capacity of the infrastructure;
- Damage to infrastructure during maintenance and repair;
- Unsafe movement of people and vehicles in the road corridors; and
- Premature deterioration of the road infrastructure.

4.2.3   Stakeholder Interviews Concerning Infrastructure Management

There are many authorities operating urban infrastructure within the road corridor. It was observed that authorities had no coordination, and each acted independently to achieve the same goal in the road corridor. Interviews disclosed how urban infrastructures were developed, managed and operated in the road corridors and their effects to infrastructure operations as well as movement. The main observations from the interview are summarised in Table 4.8. Interviewed officials included the MoID, MoWI, PMOLAG, MoHA, MoE, TANROADS and TTCL. Others were from TANESCO, DAWASA, DAWASCO, Traffic Police, City and Municipal Councils.

As found out authorities managing the city road network did not have enough knowledge about the utility network laid in the road corridors. Planning, designing, implementation and management of operations were being done separately without communication among operators. Some operators were consulted only if their networks were an obstacle during operations. Development and management of

the roadway were not synchronised with development and management of utility network in the road corridor. Such as unsynchronised activities in the corridors adversely affected movement and safety.
The observed lack of coordination among the operators was reported to be source of the major problems found in the road corridors. The road authorities complained on prolonged repair of underground infrastructure, which included cutting the road surfaces and left the cuts for a long time unprotected to the road users. Failed cut road sections and lack of maintenance to hazardous urban infrastructure were found to affect movement and safety.

The operators acknowledged that not all underground networks in the road corridors were known to the operators. Many underground utilities placed under the roads were laid during colonial era, between 1958 and 1960. Also no designs of the layout were left for reference. The repairs to the network were implemented in ad-hock style depending on where failures were reported.

The networks were mainly affected by road construction equipment and were also affected by one another as one operator repairing their network cut other services since they lacked proper knowledge of the underground layout. They were accidentally cut since the owner did not know where they were laid. Such kind of operations was revealed to affect movement and safety. Also posed potential risks to vehicle manoeuvres.

Table 4.8
Main observations from the stakeholders interview

Qns	Infrastructure	Responses
1	Physical layout and condition	There is no joint master plan of urban infrastructure in Dar es Salaam. The individual master plan available has never been timely implemented.
		There is lack of knowledge of the underground infrastructure network laid under the road corridor, among the UI operators. The exact location and layout of underground utilities are not clearly known to infrastructure operators.
2	Operations and management	Uncoordinated urban infrastructure operations, inferior maintenance techniques and defects in the urban infrastructure in the road corridor have been responsible for deteriorated infrastructure and safety.
		Repairs are implemented on an ad-hoc basis-only when leakages and bursts are reported.
		There is no urban infrastructure management system in the road corridor.
		There are inferior planning, inferior data base coupled with uncoordinated planning, development, implementation and lead to affecting traffic movement and safety
3	Monitoring and evaluation	There is neither a follow up system in place for monitoring and evaluation nor an institution responsible for coordinating activities and regulating the road corridors.
		The urban infrastructure operators are experienced with high cost of project implementation and high urban infrastructure operation cost.
		Repair and extension of the underground utilities has imposed severe damages to the road surface, damage of urban infrastructure network to one another and prolonged incidences which interrupted smooth of the traffic flow and lower the capacity of the road network.

Source: Field data survey (2007)

Stakeholders were of the opinion that there was a need to have a coordinating agency so as to manage the road corridor, coordinate operations in the road corridor and have a joint master plan to ensure synchronisation of activities in the urban road corridor.

4.3   Analysis

4.3.1 Hypotheses Testing

In responding to the first objective, two scenarios were identified where linkage was established among the infrastructure assets. Thus the study established the association between proximity and fault interactions. With such a scenario, a correlation analysis was used to test the association or correlation.

Again, the same objective sought to find out the relationship between infrastructure assets and average safety of movement. To achieve this, linear regression analysis relationship between fault interactions and average safety of movement ratings and fault interactions were developed. Moreover, the relationship between the ratings for the design and maintenance including average safety rating movements was developed.

Null Hypothesis: There is no association between road proximity and fault interactions of the
                                          road corridor.
Alternative Hypothesis: There is association between road proximity and fault interactions of the
                                          road corridor.

4.3.2   Correlation Analysis between Proximity and Fault Interactions

Correlation analysis between proximity and fault interactions of the corridor was run at 1 percent level of significance. Looking at the output of the correlation analysis, it was observed that there was correlation between proximity and fault interactions. Table 4.9 presents the correlation between the two. Results were significant, inferring that the null hypothesis was rejected, showing there is association between the two.

Moreover, running the two variables to check whether or not proximity influenced fault interactions, a linear regression analysis was conducted.

Table 4.9
Correlation between proximity and fault interaction

		Promity level of corridors	Average Effects ratings
Promity level of corridors	Pearson Correlation	1	.361(**)
	Sig. (2-tailed)		.000
	N	324	120
Average Effects ratings	Pearson Correlation	.361(**)	1
	Sig. (2-tailed)	.000	
	N	120	120

** Correlation is significant at the 0.01 level (2-tailed).

Source: Field data survey (2008)

Looking at the model test of the regression, the data deserved the test as shown in Table 4.10 as it was significant. It implies that the null hypothesis was rejected at 5 percent level of significance.

Moreover, looking at the type of relationship, the following were deduced: R=36.1 percent shows a weak positive relationship between proximity and fault interactions. It means that as proximity approached 1, more fault interactions are likely to happen. However, the adjusted R^2 =12.3 percent reflects that 12.3 percent of the fault interactions were caused by proximity variable. Table 4.10 presents the detailed analysis.

Table 4.10
Analysis of variances (ANOVA (b))

Model	Sum of Squares	df	Mean Square	F	Sig.
1 Regression	13.906	1	13.906	17.674	.000(a)
Residual	92.846	118	.787		
Total	106.752	119			

a Predictors: (Constant), Proximity level of corridors
b Dependent Variable: Average Effects (Fault Interactions) ratings
Source: Model results (2008)

Furthermore, the coefficient and the constant of the relationship were observed to be significant. Thus, a linear relationship can be stated as follows:
Number of fault interactions in a link= 0.743 Proximity level of the link + 2.8
More details are presented in Table 4.11.

The urban infrastructures are geographically interdependent in terms of sharing location in the road corridor. They are placed randomly in the road corridor in a way that no one can work on one underground infrastructure without affecting the others.

Table 4.11
Analysis of coefficients (a)

Model	Unstandardized Coefficients		Standardized Coefficients	t	Sig.
	B	Std. Error	Beta	B	Std. Error
1 (Constant)	2.800	.148		18.940	.000
Proximity level of corridors	.743	.177	.361	4.204	.000

A Dependent Variable: Average Effects ratings
Source: Model results (2008)

In order to reverse the trend, integrated high performance infrastructure has to be adopted. It will bring about the best practice through coordinated planning, design implementation, management and operations of urban infrastructure in the road corridor. Through best practices, the underground utility crossings (ducts) and utility corridor for underground infrastructure passages would be provided through integrated infrastructure. That will minimise cutting of the road surface and hence, retain good road surface condition. Incidences in the road corridor would be minimised which would also minimise the potential safety risks and conflict.

4.4 Conclusion

With increased frequency of urban infrastructure operations in the road corridors, their effect and conflict in the road corridor have been significant. The severe effect of one infrastructure to the other and potential conflicts which lead to affecting movement and safety in the road corridor were also observed. The operators reported that conflict of UI network is number one construction problem to the users in the road corridor. As more and more UI assets are frequently breaking down in the underground of the road corridor, the opportunity for these costly conflicts continues to grow.

The costs associated with the mitigation of conflicts after construction begins, include not just construction change orders, but serious delays and service disruptions to the public as well as effect on movement and safety. The relocation to the utilities also costs the developer highly.

Unfortunately, as mentioned in a preceding section there is no coordination among the operators. The urban infrastructure network operations have mainly been a reactive undertaking, responding to the reported breakdown, rather than a proactive process that begins at project conception. It was proposed by urban infrastructure operators that the proper planning, locating and coordination between involved stakeholders will minimise costs, delays, conflicts and potential hazards which are currently affecting movement and safety.

There is relative displacement of each other in the available spaces within the corridor. This is the tendency of disturbing presence or working environment of each other within the corridor when one utility is subjected to operational tasks by displacement of more spaces that would otherwise be available to all. For example, repairing of fault of underground cables can require detouring of traffic or pedestrians from their normal passages.

By the existence of supplier and customer relationship within the service providers, interruption of services supplied by one provider affects service delivery level by others. For example, electricity interruption results into malfunctioning of traffic lights.

The analysis revealed that the null hypothesis was rejected at 1 percent significant level and accepted the alternative hypothesis that infrastructure assets (such as the roadway, utilities and other facilities) in an urban corridor are geographically interdependent primarily by virtue of physical proximity as well as operational and fault interactions among the systems.

In order to contain negative interactions among infrastructures, an integrated high performance infrastructure has to be adopted. It will bring about the best practice through coordinated planning, design implementation, management and operations of urban infrastructure in the road corridors. Through best practices, the underground utility crossing (ducts) and utility corridors for underground infrastructure passages will be provided through integrated infrastructure. That will minimise cutting of the road surface and hence, retain good road surface condition. Incidences in the road corridors will be minimised which will also minimise potential risks and conflict associated with interactions of non integrated infrastructures, thereby enhance movement and safety. Results from the analysis provide evidence not to reject the hypothesis that there is association between proximity and urban infrastructure fault interactions in the road corridor.

5

5.0   MOVEMENT AND SAFETY IN ROAD CORRIDORS IN DAR ES SALAAM

5.1   Introduction

This chapter presents a discussion on overall level of movement and safety situation in the study area. It presents the relationships between the road corridor environment on the one side and movement and safety on the other. It is based on the assessment of elements of corridor environment and the extent to which such elements influence the level of movement and safety within the road corridors in Dar es Salaam.

As deduced from the preceding chapter, road corridors in Dar es Salaam were found to experience undesirable interactions of urban infrastructure. Existence of such interactions was seen to be accompanied with deterioration in level of movement and safety within the corresponding roadway. However, it was unclear whether there was any association between the extent of interaction of urban infrastructure and the noted affected level of movement and safety.

It is vital to understand the existence of the relationship between events of underground infrastructure interaction and impact on movement and safety because degradation in level of movement and safety is burdensome to the society as pointed out by Mohan (2002). Frequently, congestion, delays and occurrence of accidents are a burden to society. As a way of expressing, the chain of such burden, the quantifiers of movement and safety have been expressed from the society's point of view in terms of impact on elements of mobility such as road traffic movement as summarised in Figure 5.1.

Events causing temporary capacity losses occur in an environment comprised of roadway characteristics, location, time, and ambient conditions. Characteristics of the event and its environment provide information based on traffic impact models that can predict impacts. Such travel impacts include delay and the four Rs: (a) re-routing, (b) re-scheduling, (c) reduced mobility (foregone travel, cancelled trips), and (d) reduced reliability. In general, delay is probably a more useful and certainly more intuitive measure of loss of functionality. Thus, an event will generate all five types of impacts with relative importance of each depending on the nature of the event and its context. For example, an unexpected event such as a crash is likely to produce relatively less re-routing and re-scheduling than a work zone whose existence can be known in advance, which may persist for days, weeks, or even longer.

Figure 5.1
Influence of level of movement and safety to society

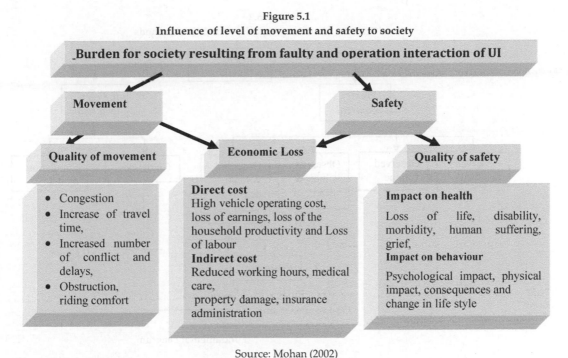

Source: Mohan (2002)

5.2   Level of Movement and Road Capacity

5.2.1   Conceptual Review of Movement within the Roadway

Movement as referred to in this study is the mobility within the road corridor. It is further conceptualised as a steady traffic stream within the stretch of road corridor. Description of movement as a measurable and quantified item is complex and it can just be nominally expressed in terms of desirable and undesirable level. However, for such level to be categorised, conditions that define levels have been inferred from parameters that are measurable. Such parameters describe the capacity of the road, which in turn reflects the level of movement within the road stretch. Therefore, as capacity reflects possible level of movement that can be attained, the main determinant factors for the capacity are flow, speed, density of traffic and level of movement.

A critical distinction is made between loss of capacity and its impact. Capacity is a measure of potential; it describes maximum sustainable through put of a highway. As such, it is independent of the road's actual use. Impacts, however, depend not only on the loss of capacity, but also on the volume of traffic on the road when the loss occurs.

Several approaches have been developed to express road capacity. Figure 5.2 presents a scheme in which various approaches are distinguished. There are direct and indirect empirical methods. Based on direct empirical approaches, Arem and Vilst (1992) determined operational capacity through dynamic capacity models such as the on-line procedures.

Figure 5.2
Classifications of roadway capacity estimation methods

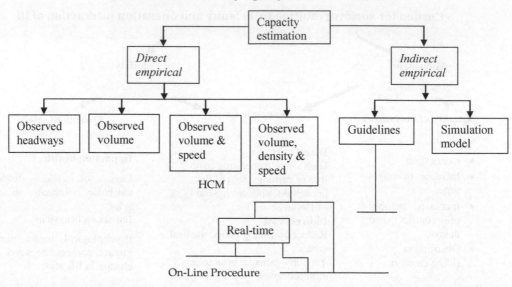

Fundamental Diagram

Source: Minderhoud, *et al.* (1992)

Yagar and Vanar (1983) listed factors affecting capacity and speed-flow relationship for two-lane highways under the following three headings;

- Geometric factors: grades, bendiness, lane width and lateral clearance;
- Traffic factors: vehicle mix, abutting land use (not really a traffic factor) and turning movements; and
- Weather-surface factors: darkness, pavement roughness and the winter season alone (without adverse weather) all decreased speed.

Also Schofield, *et al* (1986) identified the following reasons why there is a range of speeds associated with a given roadway flow:
- Vehicle composition.
- Weather: highest speeds were observed in dry and clear conditions. Precipitation was found to reduce speeds. Dry clear conditions were associated with speeds 6-20 km/hr higher than wet conditions.
- Light conditions: the speed difference between daylight and darkness was found to be small at low flows but increased with increasing flow: at 2000 veh/h/carriageway the difference was about 2 km/h, but at 5000 veh/h it was 10 km/h or more. Schofield conducted this research on lighted roads, and he suggested that the effect would be greater on unlit roads.

The generalised relationships between speed, flow and density of traffic, which are the basis for the capacity analysis of uninterrupted-flow facilities, are presented in the American Highway Capacity Manual 2000 (HCM) as shown in Figure 5.3. The importance of understanding the relationship between flow, speed and density is unquestionable. From the standpoint of design, knowledge of high flow rate characteristics is required for the prediction of road capacity. Also from the standpoint of traffic operations, understanding the entire range of relationships is important to provide adequate level of service.

Figure 5.3

Generalised relationships among speed, density, and flow rate on uninterrupted-flow

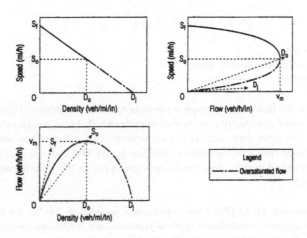

Source: HCM (2000).

Based on definition from the HCM (2000), capacity is defined as "the maximum hourly rate at which persons or vehicles can reasonably be expected to traverse a point or uniform section of a lane or roadway during a given time period under prevailing roadway, traffic and control conditions." The time mostly used in the HCM (2000) is 15 minutes, which is considered to be the shortest interval during which stable flow exist.

Density (K) also called concentration, is defined by HCM (2000) as "the number of vehicles occupying a given length of lane or roadway, averaged over time, usually expressed as vehicles per unit distance." Freeways or motorways are the only type of roads assumed by HCM (2000) to be completely uninterrupted. All other road types experience a greater or lesser degree of interruption. HCM (2000) explains that though form of relationships is the same for all such facilities, the exact shapes and their numeric calibration depend on the traffic as well as roadway conditions of the highway under study.

The curves in Figure 5.3 illustrate several significant points. First, a zero flow rate occur under two different conditions. One is when there are no vehicles on the facility, density is zero and flow rate is zero. Speed is theoretical for this condition and would be selected by the first driver (presumably at a high value). Such speed is represented by V_f in the graphs. The second is when density becomes so high that all vehicles must stop, the speed is zero and the flow rate is zero because there is no movement and vehicles cannot pass a point on the roadway. The density at which all movement stops is called jam density, denoted by K_j in the diagrams. Between these two extreme points, dynamics of traffic flow produce a maximising effect. As flow increases from zero, density also increases, since more vehicles are on the roadway. When this happens, speed declines because of the interaction of vehicles. Such decline is negligible at low and medium densities and flow rates. As density increases, these generalised curves suggest that speed decrease significantly before capacity is achieved. Capacity is reached when the product of density and speed results in the maximum flow rate. Such condition is shown as optimum speed V_0 (often called critical speed), optimum density K_0 (sometimes referred to as critical density) and maximum flow q_m.

The slope of any ray line drawn from the origin of the speed-flow curve to any point on the curve represents density, based on this equation:

$$K = Q/V \qquad\qquad\qquad (5.1)$$

where:
    Q = flow rate (veh/hr)
     V = average travel speed (km/hr), and
    K = density (Veh/km)

Similarly, a ray line in the flow-density graph represents speed. As examples, Figure 5.3 shows average free-flow speed (V_f), speed at capacity (V_0), as well as optimum (K_0) and jam (K_j) densities. As shown in Figure 5.3, any flow rate other than capacity can occur under two different conditions, one with a high speed and low density, second with high density and low speed. The high-density, low-speed side of the curves represent oversaturated flow. Sudden changes can occur in the state of traffic (that is in speed, density and flow rate).

Through empirical research, HCM (2000) has documented speed-flow models for typical uninterrupted and interrupted-flow segments on different types of facilities, that is freeways, multilane highways, two-lane highways and urban arterial streets of different classes as shown in Figure 5.4. It is indicated that different road facilities reflect different speed-flow characteristics. Uninterrupted facilities are usually characterised by high Free-Flow Speeds (FFV), which are sustained at the same level during low volume traffic flows until near capacity, while interrupted facilities are characterised by low Free-Flow Speed (FFV) whereby travel speeds are sensitive to traffic flow volumes and other variables including signal density and urban street class.

The HCM (2000) methodology for analysing urban streets does not adequately address some conditions that can occur between intersections, which are potential sources of changing traffic flow. One of such factors is interruptions caused by interactions of urban infrastructures.

The referred literature lacks recognition of the impact of faults and operational interactions of urban infrastructure network into the level of services for the roads. Such gaps are addressed in the following sub-sections so as to understand the contribution of interactions into level of movement within Dar es Salaam road corridors.

Figure 5.4
An interrupted flow

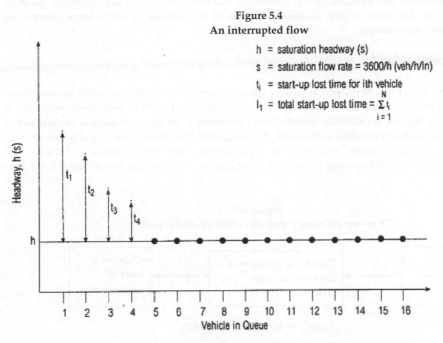

h = saturation headway (s)

s = saturation flow rate = 3600/h (veh/h/ln)

t_i = start-up lost time for ith vehicle

l_1 = total start-up lost time = $\sum_{i=1}^{N} t_i$

Source: High Capacity Mannual (2000)

5.2.2   Movement Evaluation along Road Corridors in Dar es Salaam

Movement, which is considered in the section, is limited to vehicles only because pedestrians and other forms of traffic have been observed to be randomly generated within the link, as such, there are many random origins and destinations localised within the links. Hence, records of such nature cannot be used to validate parameters required for determining the level of movement within the stretch of interest. In order to explain the extent of effects of urban infrastructure interaction on movements within the study corridor, operational capacity was adopted as an explanatory variable for showing the level of movement. Operational capacity referred to here is a capacity value representing an actual maximum flow rate of the roadway, which is assumed to be a useful value for short-term traffic forecasting with which traffic control procedures may be performed. The researcher adopted the model, which is based on approaches of HCM (2000) to study the extent of influence of the urban infrastructure interaction on movement within the roadway.

Survey data collected by real time records using CCTV cameras within the links that exhibited similar conditions related to the following assumptions:
- Having uninterrupted flow from the access points, and
- Having homogeneous geometric characteristics.

Data were collected from each of the study roads under two scenarios as follows:
- In situations of no faults or operational interactions among existing infrastructures, and
- In events of occurrence faults or operational interactions among the existing infrastructures.

The collected real time data were analysed to deduce whether or not there were any significant statistical evidence to infer that there existed significant impacts of interactions of urban infrastructures on movement within the roadway.

5.2.3   Conceptual Model

A statistical test model was developed to analyse whether interactions among the urban infrastructures had significant influence on the road capacity, which, in turn affects level of movement. Basically, an average capacity per lane on different corridor was considered to be equal, regardless of existence of incumbent infrastructure interactions. However, it is suggested that the average capacity per carriage way differs significantly under circumstances of occurrence and non-occurrence of Faults and Operational Interactions (FOI) among the urban infrastructures. This concept is summarised in Figure 5.5.

Figure 5.5
Conceptual model of effect of FOI on road capacity

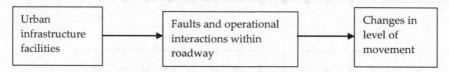

Source: Field data survey (2007)

The adopted procedure for testing equality of several means is the variance analysis. Variance analysis was applied for single factor to model the effect of the occurrence of FOI on road capacity.

Let C_{ij} be the j-th observation of average capacity per lane under i-th event of with or without FOI. Then, the linear statistical model of average capacity per lane on different highways can be written as:

$$C_{ij} = c + \tau_{ij} + \varepsilon_{ij} \quad \text{Where} \quad \begin{cases} i = 1, 2, 3, \ldots . m \\ j = 1, 2, 3, \ldots . n \end{cases} \tag{5.2}$$

Where c is a parameter common to all treatments called overall mean of lane capacity on different highways, τ_i is a parameter unique to the i-th treatment called the i-th treatment effect, that is, the random effect on i-lane highway, and ξ_{ij} is a random error component.

Objectives of the study were to test appropriate hypotheses about average capacity per lane on different highways under both situations with FOI and without FOI. For hypothesis testing, model errors are assumed to be normally and independently distributed random variables with mean zero and variance σ^2. The variance σ^2 is assumed to be constant for all levels of the factor.

Let C_i be average capacity per lane on i-th event of with or without FOI, and

$$C_i = C + \tau_i \; ; \text{ for } i=1, 2, 3 \ldots .m \tag{5.3}$$

Thus, the mean on i-th event consists of the overall mean plus the random effect on i-th event.

The null hypothesis and alternative hypothesis can be expressed as:

$H_0 : C_1 = C_2 = C_3 = C$

$H_1 : C_i \neq C_j$ for all value of i, j, and $j \neq i$.   (5.4)

Note that if H_0 is true, average capacity per lane with and without FOI has common mean C.

The null hypothesis assumes that road capacity is indifferent under both situations of with and without FOI. That is to say, is not different in average capacities per lane on highways with and without FOI $(c_1 = c_2 = c)$. Then, the variance analysis of single factor was applied to test the null hypothesis.

Finally, one-sided t-test of independent samples was used to test the decrease of capacity as the events of FOI increases c_i and c_{i+1}, In these problems, H_0 would be rejected only if one mean is larger than the other. Thus, their null hypotheses and alternative hypotheses can be written as:

$H_0: c_1 = c_2$, $H_1: c_1 > c_2;$ (5.5)

$H_0: c_2 = c_3,$ $H_1: c_2 > c_3;$ (5.6)

If both of null hypotheses are not true, there exists a gradual decrease of c_1, c_2 and c_3

5.2.4   Data Collection through Survey

Locations of surveyed areas are as shown in Table 5.1. Appropriate links for observation were chosen within roads based on historical information, which pointed out particular links prone to events of FOI and with potential of being extended to cover study duration.

Capacities on these links were negligibly interrupted by other factors except on events of occurrence of FOI of urban infrastructure. Table 5.1 summarises information related to chosen locations within the studied roads.

The survey period and characteristics of roads within the studied locations are as shown in Table 5.2. From this table geometric characteristic such as lane width, lateral clearance, and design speed on different highway sections are consistent on all road sections.

Table 5.1

Summary of studied road links

Road Name	Location of study link	No. of lanes	Registered FOI events
Bagamoyo	Ohio	2	Damaged manhole cover, overflow on sewer, road surface failure
	Ali Hasani Mwinyi	4	Burst of water pipe, overflow of sewer
	New Bagamoyo	3	Open manhole, leakages of water pipes
Morogoro	City Drive-NSSF	2	Damaged manhole, cut road surface for underground repair of sewer system
	Fire-Magomeni	4	Water layer on the road surface, burst of water pipe, leakage of sewer pipe
	Magomeni-Ubungo	4	Damaged manhole, overflowing of sewer
Nyerere	Ohio-Uhuru	4	Protruded manhole on the road surface, overflowing of sewer on the road surface
	Uhuru – Changombe	4	Burst of water pipe, damaged manhole
	Changombe-TAZARA	4	Overflowing of water on the road surface, leakage of sewer pipe
Kilwa	Kivukoni-Railway	2	Depression of road surface due to underground infrastructure failure, leakages of sewer pipes
	Railway-Bandari	4	Damaged manhole, damaged road surface
	Bandari-Uhasibu	4	Relocation of utilities, overflowing of sewage and water due to destruction of water pipes and sewer.
Mandela	Mwenge-Ubungo	4	Overflowing of water pipe mixed with high tension electrical cable
	Ubungo-TAZARA	4	Damaged manhole, damaged road surface
	TAZARA-Bandari	4	Overflowing of water in the road surface, cut of road surface for underground infrastructure repair

Source: Field data survey (2007)

Entrance, exit or curvature is over seven hundred meters from each surveyed location. Therefore, geometric characteristics have little influence upon highway capacity. Furthermore, weather conditions during the survey were clear. Thus weather influenced capacity on surveyed roads.

Table 5.2

Summary of field capacity survey

Common characteristics of road sections	a) lane width; 3.25 – 3.7 m, shoulder 1 – 2 m
	b) sight distance > 300 m (no restriction on overtaking)
	c) road surface: good condition
Date and time	date: (workdays) Monday to Saturday
	time: 0700 to 1000 hrs
Number of samples	Six for each of both cases of with and without FOI in all five roads

Source: Field data survey (2008)

Finally, all surveys were done on workdays from 7.00 to 10.00 a.m. Activity types of travellers are mostly work trips. Data on spot flows, speeds, densities and vehicle classifications by lane were collected in one direction over 15 minutes categorised into a group. Figure 5.6 shows photographs of surveyed road sections.

Figure 5.6
Photographs of surveyed road sections

Source: Field data survey (2007)

Classification of vehicle types was based on three vehicle-length classes of up to less than 5 m, 5 m to 8 m, and more than 8 m. All vehicles are classified into passenger car, medium vehicle and heavy vehicle and their passenger car equivalents are 1.0, 1.5 and 2 m, respectively. The detailed vehicle classification for road capacity analysis is shown in Table 5.3.

Table 5.3
Vehicle classification for road capacity analysis

Type of vehicle	Description of the vehicle type	Length of vehicle (m)
Passenger car	Mini-buses, jeeps, vans, passenger cars, light vehicles, and microbuses(<12 p), etc.	< 5
Medium vehicle	Medium bus(12-25 p), trucks(2.5-7 t), etc.	5 to 8
Large vehicle	Large trucks(>7 t) large buses(>25 p), trailers, heavy vehicles, and container-vehicles, etc.	>8

Source: Field data survey (2007)

5.2.5   Survey Results

On the basis of Table 5.4, percentages of vehicle type can be obtained. The percentage of medium vehicles on each road was lower than 4 percent and differences between them were less than 1 percent. Meanwhile, the percentage of heavy vehicles was very small, less than 0.1 percent on each road. It implies that differences to percentages of each vehicle type on different roads are different. The passenger car equivalents might lead to little error of road capacity.

Table 5.4
Passenger car equivalent in the road corridor along Morogoro Road

Time Interval	Passenger car	Medium vehicle	Heavy vehicle	Total
06:00 – 06:15	67	14	6	87
06:15 – 06:30	83	18	8	109
06:30 – 06:45	106	21	11	138
06:45 – 07:00	139	25	12	176
07:00 – 07:15	184	36	14	234
07:15 – 07:30	258	51	19	328
07:30 – 07:45	237	46	16	299
07:45 – 08:00	211	41	14	266
08:00 – 08:15	193	38	13	244
08:15 – 08:30	178	35	12	225
08:30 – 08:45	161	32	14	207
08:45 – 09:00	247	48	18	313
09:00 – 09:15	220	43	15	278
09:15 – 09:30	186	36	14	236
09:30 – 09:45	164	32	12	208
09:45 – 10:00	142	28	11	181
Total	2776	544	209	3529
Percentage (%)	79	15	6	

Source: Field data survey (2007)

5.3   Average Capacity per Lane on Highways

The American Highway Capacity Manual (1994) defines highway capacity as: "The maximum sustained 15-minute rate of flow which can be accommodated by a uniform highway segment under prevailing and roadway conditions in the specified direction of interest". Therefore, 15-minute flow rate was used to assess road capacity.

Table 5.5
Summary of average capacity per lane on each road (PCU//h)

ROAD NAME	STATUS	Monday to Saturday						
		Mon	Tue	Wed	Thur	Frid	Sat	Average
Morogoro Road	Without FOI	1003	945	997	1122	1045	750	977
	With FOI	800	766	781	825	819	659	775
Nyerere Road	Without FOI	669	630	665	748	697	500	652
	With FOI	533	510	521	550	546	439	512
Kilwa Road	Without FOI	514	485	511	575	536	385	501
	With FOI	410	393	400	423	420	338	397
Mandela Road	Without FOI	720	679	716	806	750	538	702
	With FOI	574	550	561	592	588	473	556
Bagamoyo Road	Without FOI	771	726	767	863	804	577	751
	With FOI	615	589	601	635	630	507	596

Source: Field data survey (2007)

Observations obtained on average capacity per carriageway on different roads are shown in Table 5.5. It is the case that the means of observation on different roads were significantly different. The means of observation on different roads gradually decreased with increase in events FOI. Meanwhile, standard deviations on different highways underwent a little change, varied from 30.79 to 34.83. However, the standard deviation of overall observation was as nearly 3.5 times as that of each road.

Figure 5.7 shows variation of average capacity per carriageway along Morogoro road. Solid dots are individual average capacity observations in a day. The recorded flow with faulty operation interaction of UI is in red, while flow without faulty operation interaction is in blue colour.

Figure 5.7
Average capacity per carriageway along Morogoro road

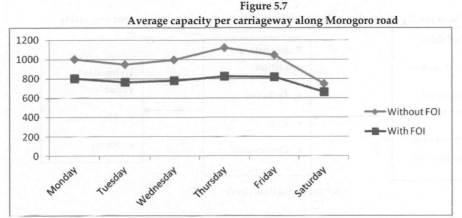

Source: Field data survey (2007).

The graph indicates that average capacity per lane decreases as FOI increases. Figure 5.7 shows average capacity per carriageway decreases with increasing number of FOI. Results from the analysis also show a statistical significant difference in average speed of different roads of each category.

Figure 5.8
Traffic flow along Morogoro Road

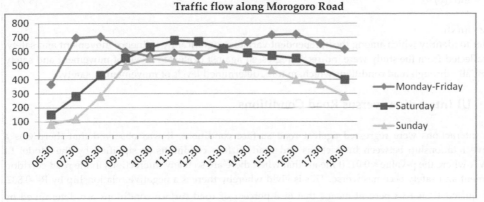

Source: Field data survey (2007).

Figure 5.8 shows recorded flow of traffic by surveillance camera from Monday to Sunday. The impact of FOI is explained by increasing speed of approximately 2 percent or 1.1 km/hr. They were found to be significant by the F-test, and hence, concluded that the impact was significant at confidence $\alpha = 0.05$ level.

5.4   Analysis of UI versus Movement and Safety in the Road Corridor

The conceptual relationship of this scenario was based on two levels of interdependencies in form of implicit functions involving UI variables, road surface condition variables and movement or safety variables as shown in Figure 5.9.

Figure 5.9
Conceptual presentation of the causal-effect between UI and movement and safety

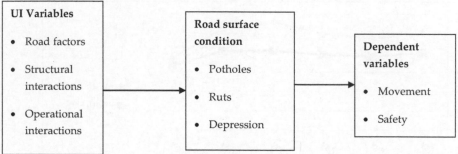

Source: Field data survey (2007)

If denoted;

x as a set of independent variables expressing UI interactions,

y as a set of dependent variables of independent variables x expressing road conditions, and

z as a set of dependent variables of variables y expressing level of movement or safety

Then, the conceptual presentation of this relationship can be expressed as

$f(x) = y$ and $f(y) = z$ (5.7)

Thus $z = f(f(x))$

In order to identify which among the independent variables significantly influence movement and safety, data collected from the study were indirectly regressed against evaluated levels of movement and safety sequentially through road conditions which, in turn, determined levels of movement or safety.

5.4.1   UI Interactions versus Road Conditions

When interactions were regressed against road surface conditions, it was observed that there was a negative relationship between interactions and road surface conditions as signified by the Model fit ANOVA where the p-value < 0.01. It means that when there are so many interactions in the road corridor, movement and safety becomes worse. This is vivid whereby there is a negative relationship by R=-0.832. The Adjusted R^1 is 68.8 percent means that 68.8 percent of road surface conditions are determined by interactions (See annex II). Moreover, coefficients proved to be significant and results are presented in the Table 5.6.

Table 5.6
Analysis of variances (ANOVA)

Model		Sum of Squares	Df	Mean Square	F	Sig.
1	Regression	42.993	3	14.331	183.780	.000[a]
	Residual	19.183	246	.078		
	Total	62.176	249			

Source: Model analysis results (2008)

Results in Table 5.7 indicate that all predictor variables were significant and the whole model explained high explanatory power to variance of criterion variable as depicted by the R^2 value and it was significant as shown by ANOVA statistics. Importantly, results identified the value of regression coefficients

Table 5.7
Analysis of coefficients[a]

Model		Unstandardized Coefficients		Standardized Coefficients	t	Sig.
		B	Std. Error	Beta		
1	(Constant)	.178	.058		3.088	.002
	Road surface interactions	.261	.068	.258	3.816	.000
	Urban physical interactions	.402	.065	.402	6.172	.000
	Functional interactions	.230	.075	.228	3.049	.003

Source: Model analysis results (2008)

5.5 Overall Safety Situation

5.5.1 How Tanzania Compares with other Countries in Road Accident

The magnitude of road accidents can be comprehended by looking at the trend of accident rates or accident frequency overtime, by comparing accident rates with other countries and by comparing road accident fatalities with other causes of death in society. Generally, the impact of road traffic accidents in terms of injuries, impairment and fatalities is still a global, social and public health challenge.

In examining the trend, apparently Tanzania has a big number of road traffic accidents and causalities. The number of accidents reported by the traffic police has tremendously increased in more than 30 years from 7,850 (in 1975) to 17,264 accidents in 2007. The total numbers of 2,366 people were killed in 1,851, fatal accidents reported by police in Tanzania in 2007. During that year, the statistics showed an alarming increase of number of accidents and number of killed as well as seriously injured persons by 7 -11 percent.

Figure 5.10 indicates the number of accidents and fatalities that occurred in Tanzania. They have been increasing with time. They reached more than 20,000 in the year 2007. The data indicate also that more than 10 percent of accident victims in each year were killed. The increase in accidents from 1990 to 2007 was 76 percent which is +4 percent annually. Fatalities increased from 1990 to 2007 were 145, +9 percent per year. Registration in Tanzania requires the driver of a vehicle involved in a personal injury accident to report the incident to the police. According to regulations, all accidents are required to be reported to the police, regardless of the severity of such accident. In practice, however, the police are notified only when an accident involves serious injury or if agreement cannot be reached between involved parties.

Figure 5.10
Road accidents victims in Tanzania 1975-2007
Road Safety Situation in Tanzania

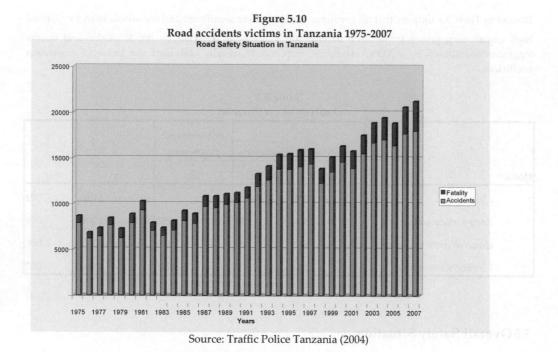

Source: Traffic Police Tanzania (2004)

Reporting of fatal and non-fatal accidents is uncertain and under-reporting of road traffic accidents in Tanzania was noted to be conspicuous. Furthermore, most reports do not contain details of road sections and precise locations of accidents reported. Location of an accident is reported broadly by using the name of an area instead of the road name and location name of the scene using proper GPS coordinate.

It is known that accidents are rarely randomly distributed along roads. Conversely, they tend to occur in clusters (Hasselberg, *et al* 2007). Therefore, it is very important that the location of an accident is accurately recorded in the accident report in order to identify precise causes of accidents.

Table 5.8
Distribution of fatalities and seriously or slightly injured persons

S/N	Road User Group	Fatalities (%)	Seriously or slightly injured persons (%)
1	Drivers	12	5
2	Passengers	43	54
3	Motorcyclists	4	3
4	Pedal Cyclists	8	7
5	Pedestrians	33	31
	Total	100	100

Source: Tanzania Traffic Police (2007)

Road traffic accidents constitute a major problem in Tanzania. The two major risk groups are passengers and pedestrians, according to the data provided in Table 5.8. The recorded fatalities for passengers and pedestrians were 43 percent and 33 percent respectively, while for injuries to the passengers and pedestrians were 54 percent and 31 percent respectively. This indicates that passengers and pedestrians are the most vulnerable road users in Tanzania.

Table 5.9
Major contributing factors of road accidents

Causes of Road Accidents	Years						Sub Total	Ranking %
	2000	2001	2002	2003	2004	2005		
Reckless driving	7,041	6,743	8,179	10,916	9,366	8,962	51,207	54.47
Defective vehicles	2,797	2,440	2,641	2,503	2,403	2,234	15,018	15.98
Careless pedestrians	650	1,056	1,096	1,463	1,337	1,248	7,050	7.5
Excessive speed	426	350	340	376	1,409	578	3,479	3.7
Careless M/cyclists	924	827	827	483	757	1,405	5,223	5.56
Careless cyclists	1,276	891	891	367	607	758	4,790	5.1
Intoxication	170	98	99	68	171	163	769	0.82
Faulty road surface	1,064	1,472	1,417	488	989	1,040	6,470	6.88

Source: Traffic Police Report (2007)

Table 5.9 illustrates the main causes of road accidents as reported by Traffic Police in Tanzania. Among such reported causes of accidents, reckless driving was dominant and contributed 54 percent, followed by defective vehicles, careless pedestrians and faulty road surface, which contributed 15.98 percent, 7.5 percent and 6.88 percent, respectively.

To get a reliable picture of road accident situation, several risk-related measures are used to estimate accident ratio and exposure. Risk is a useful measure for comparing traffic safety between road types and road user categories.

This definition can be mathematically expressed as:

$$Risk = \frac{Accident\ events}{Exposure} \qquad (5.8)$$

In the context of international comparison, accidents where accidental events had fatal consequences are normally used because fatality statistics are fairly well reported over time in many countries. Normally, traffic accidents are compared as a health risk in line with diseases. In this risk concept, population is used as a measure of exposure.

Health risk can be defined as:

$$Health\ risk \quad = \quad \frac{Fatalities}{Population} \tag{5.9}$$

In the absence of a good measure of traffic activity such as vehicle-kilometres travelled, a substitute measure of exposure, the number of vehicles, is normally used to make comparisons in traffic safety between countries.

The risk in relation to the number of vehicles can be defined as:

$$Traffic\ system\ risk \quad = \quad \frac{Fatalities}{Motor\ vehicles} \tag{5.10}$$

These rates agree with values that have been quoted in different papers by Jacob and Sayer (1983), Downing *et al.* (1991), Silcok (1991), Jorgensen (1997) and Thomas (2000).

Table 5.10
Distribution of road deaths, motor vehicles and population

Region	Fatalities	Motor Vehicles	Population
Sub-Saharan Africa	10 %	4%	10%
Developed World	14%	60%	15%
Asia/Pacific	44%	16%	54%
Central & Eastern Europe	12%	6%	7%
Latin America/Caribbean	13%	14%	8%
Middle East/North Africa	7%	2%	5%

Source: Jacobs and Thomas (2000)

About 10 percent of global deaths occur in Africa, which is slightly less than those for the entire developed world or for the whole of Latin America, Central America and the Caribbean Countries, as reported by Jacobs and Aeron-Thomas, (2000). The regional share of fatalities, population and motor vehicles, worldwide can be seen in Table 5.10.

It can be seen that whilst about 10 percent of global road deaths took place in 1999 in Sub-Saharan Africa, only 4 percent of global vehicles are registered in the region. Conversely, only 14 percent of road deaths occurred in the entire developed world (North America, Western Europe, Australia and Japan). Yet this particular region contains 60 percent of all globally registered vehicles.

Fatality rates in Eastern African countries lie in the range of 70 to 250, with Tanzania being the second from the lowest as shown in Figure 5.11. Fatality risk, the reported number of deaths per 100,000 populations is the indicator most commonly used by the World Health Organisation to prioritise diseases and other causes of death.

In East Africa, the fatality risk lies between zero for Madagascar and 12 for Malawi. The comparison of Tanzania and nearby countries in terms of killed persons per ten thousands vehicles and fatalities per 100,000 populations are shown in Figure 5.11 and Figure 5.12, respectively.

Figure 5.11
East Africa Fatality Rates per Vehicle

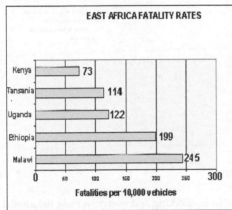

Figure 5.12
East Africa Fatality Risks per Population

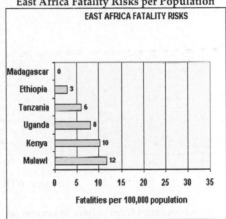

Source: Jacobs and Aeron-Thomas (2000)   Source: Jacobs and Aeron-Thomas (2000)

Comparison of road traffic accidents between countries varies due to several factors such as the reliability of accident data, differences in the way data are recorded and differences in the definition of accident fatalities. Figure 5.13 clearly illustrates a wide difference in road accident fatalities between different continents of the world (WHO, 2004). Developing countries tend to have much higher fatality rates than developed countries, which have relatively low fatality rates that lie at the lowest edge of the graph.

Figure 5.13
Percentage change in accidents fatalities

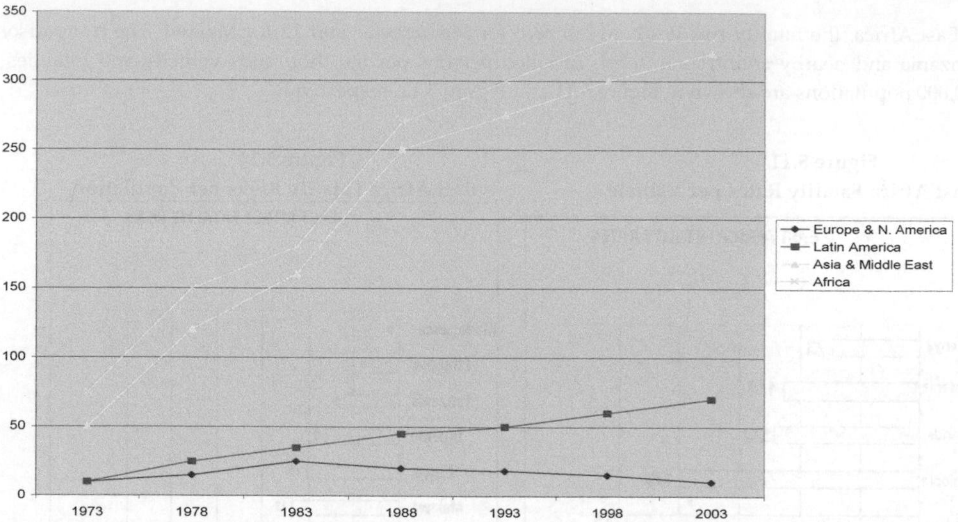

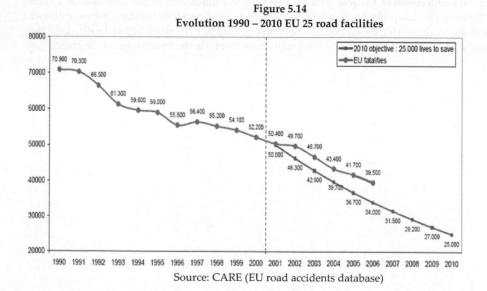

Source: WHO (2004)

While developed countries seem to have in general succeeded in checking and even reversing the annual number of road accident fatalities as shown in Figure 5.15, the number of fatalities in developing countries is seriously increasing, thus putting a heavy burden on already overloaded medical facilities and services (WHO, 2004). The fatalities in these countries have trebled from 1980 to 2010 and they will continue to increase with growing motorisation if no effective remedial actions are taken.

Figure 5.14
Evolution 1990 – 2010 EU 25 road facilities

Source: CARE (EU road accidents database)

Safety of movement of a road corridor is reflected by the rate of road accidents. Occurrences of road accidents are indicators of safety of movement vulnerability. In study locations, data of road accidents present the safety risk and the same is linked to the increased movements brought about by uncoordinated infrastructure systems.

5.5.2   Road Accidents in Dar es Salaam

According to accidents data collected, it was established that 65 percent of total accidents occurring in Tanzania occur in Dar es Salaam. Figure 5.15 shows occurrences of accidents in the country as a whole compared to accidents which occurred in Dar es Salaam.

According to Traffic Police Report (2007), Dar es Salaam has a higher record of accidents than any city and municipalities in Tanzania. This is mainly because Dar es Salaam is the main business city in Tanzania and has high interaction of urban infrastructure and traffic. There are also uncoordinated operations of infrastructure network that cause safety problems.

Figure 5.15
Road traffic accidents in Dar es Salaam and in Tanzania

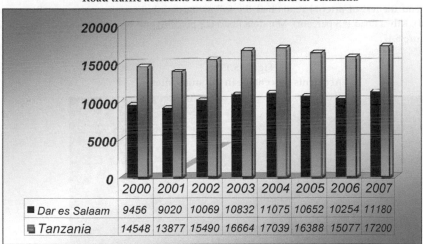

	2000	2001	2002	2003	2004	2005	2006	2007
■ Dar es Salaam	9456	9020	10069	10832	11075	10652	10254	11180
▨ Tanzania	14548	13877	15490	16664	17039	16388	15077	17200

Source: Tanzania Traffic Police (2007)

(i)   Road Accidents in the Study Corridors
Road accidents in the study corridors were analysed in the selected 49 links. Most collected accident data did not indicate infrastructure related factors, which contributed to accident occurrences. Instead, causes of road accidents were broadly reported, for example, drinking and driving, careless driving, over speeding. In rare cases, the faulty road surfaces were reported. In this case, faulty road surfaces contributed to 6.90 percent of the total number of accidents from year 2000 to year 2007 (Figure 5.15). It was revealed from interviews with Ministry of Home Affairs' officials that Traffic Police do not record accident data as required due to inadequate training and equipment. They understandably see their roles as primarily to find out if the law was broken and try to get evidence for prosecution rather than investigate various factors involved in accidents.

Figure 5.16
Road accidents indicating faulty road surface

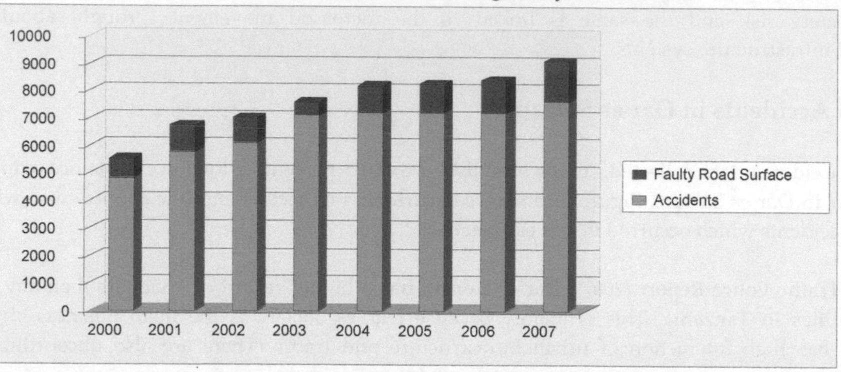

Source: Tanzania Traffic Police (2007)

According to Traffic Police records, arterial roads had a higher accident occurrence rate and severities than others. It was ascertained that 39 percent of accidents in the past eight years occurred along Morogoro road and the rest of arterial roads had the following distribution of occurrences: Bagamoyo Road (26 percent), Kilwa Road (13 percent), Mandela Road (12 percent) and Nyerere Road (10 percent) as illustrated in Figure 5.17.

Figure 5.17
Accidents percentage in the study sites

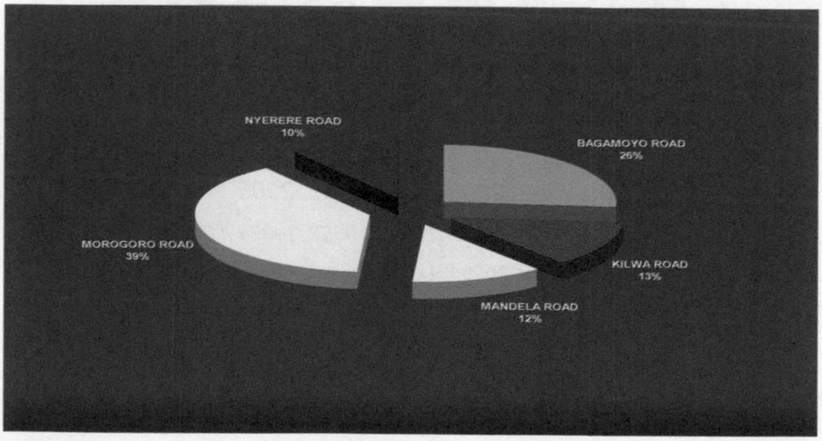

Source: Based on MAAP for windows accident analysis

The distribution and trend of the road accidents in the five road corridors are shown in Figure 5.18.
Accident distribution relative to study locations is attributed to intensity of infrastructure networks in the road corridor. Morogoro road traverses across the most densely populated residential areas such as Magomeni, Kagera, Manzese and Ubungo. It is used by both domestic and up-country traffic. On the other hand, Bagamoyo road, Mandela road and Kilwa road are located in low populated residential areas while Nyerere road mostly serves industrial areas in the city.

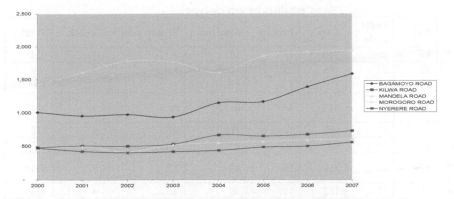

Source: Author based on MAAP for windows accident analysis

In order to establish the effect of interaction of traffic and relative movement as a cause of traffic accidents in the city, observation was further made on types of accidents that occurred in road corridors. The types were based upon vehicle manoeuvres leading to conflict situations that resulted into accidents. It was established that pedestrian accidents were the most frequent type. They constituted about 19.1 percent of all accidents, which occurred on the arterial roads in the study area, followed by head-on, sideswipe accidents and rear end accidents, with 17.4 percent, 17.0 percent and 15.1 percent respectively. It can also be noted that overturning accidents constituted about 4 percent.

The rainy season in Dar es Salaam is between February and May. These months appear to have a high rate of accidents compared with dry season months. Therefore, the effect of slippery and flooding on road corridors during wet periods can be considered to be among the major causes of accidents. Wet weather crashes represent typically 20-30 percent of road traffic accidents, and most involve skidding (Ogden, 1996).

Table 5.11
Casualties by accident type along the corridor

Accident Types	Accidents	Casualties per accident by severity			
		Casualties/accident	Fatalities	Serious Injuries	Minor Injuries
Single vehicle (loss control, poles)	533	2.26	0.45	1.07	0.74
Hit pedestrian	396	1.24	0.36	0.59	0.19
Hit other road user	232	1.49	0.36	0.76	0.37
Vehicle driving in same direction	117	2.33	0.41	1.26	0.66
Vehicle in turning/crossing direction	64	2.39	0.37	1.25	0.77
Vehicle driving in opposite direction	69	5.69	1.37	2.7	1.63

Source: Model analysis results (2008).

The number of casualties by accident type in the corridor is shown in Table 5.11 and the number of casualties by vehicle type is described in Table 5.12. There are two categories of vehicles dominating, which are minibuses and pickups. They are also the most common vehicles operating in the corridors.

Table 5.12
Casualties by vehicle type

	Bus	Pedal cycle	Cars	Pickup	Minibus	Motorcycle	Truck	Total
Fatal (n)	95	25	140	173	183	6	12	624
(ra)	52	25	105	118	108	6	7	506
(c)	2	1	1.3	1.5	1.8	1	1.7	1.4
Serious Injuries (n)	257	18	443	315	317	83	9	1442
(ra)	38	11	182	144	114	27	8	526
(c)	3.8	1.6	2.4	2.2	2.8	3.1	1.1	2.7
Minor Injuries (n)	195	16	259	259	108	12	5	854
(ra)	51	14	162	147	46	10	4	434
(c)	3.8	1.1	1.6	1.8	2.3	1.2	1.2	2
Total (n)	557	59	899	747	531	101	26	2920
(ra)	141	50	499	409	218	43	19	1379
(c)	4	1.2	1.8	1.8	2.4	2.3	1.4	2.1

Source: Model analysis results (2008)

The first row for each category of severity in Table 5.12 indicates the number of casualties (n), the next is about number of respective accidents (ra), and the last one is about casualties per respective accident (c).

5.6 Movement and Safety Hazards

5.6.1 Design-related Movement and Safety Hazards

There are several designs related to movement and safety hazards observed in road corridors of the study area. When an assessment was carried out on underground and roadside infrastructure road way operations, it was found out that safe design principles have not consistently been applied such that they endanger safety of movement. On the roadway design, consideration was on geometric features such as median, shoulder width, carriageway width, gradients, horizontal alignments and vertical alignments. Another feature observed under this category was the presence and functioning of the drainage system. On the underground infrastructure, considered features included the depth assets placed from the road surface and technical as well as equipment standards set for guidance in fixing urban infrastructure assets. On the roadside infrastructure, the observed factor was distance at which features were placed from the road edge.

The design related elements of safety hazards are minimised when the roadway, underground and roadside features are properly designed as well as located. The observed design related to defect in the road corridors under the study included shallow depth of the underground infrastructure. Contrary to the standards, most of them were located at about 0.6 m below the road surface. Consequently, underground infrastructures are vulnerable to damage. When damaged, they affect the movement and safety. Manholes observed were also not properly fixed or were found to be fixed at different levels with the road surface, which impose safety hazards to road users.

Most of the roadside features such as poles (electric power and telecommunications poles) and road-sign posts are placed at short distances between 1 to 2 metres from the edge of the road. Thus they become potential safety hazards (increase the possibility of being crashed or involved in the road traffic accident). Therefore, they are prone to damage and increase the severity of accidents. The distance of the roadside features from the edge of the road contribute a lot to the safety situation of the road users. Figure 5.19 illustrates how road accidents involving road side objects vary with the distance from the road edge.

Figure 5.19

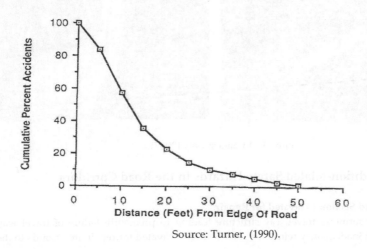

Source: Turner, (1990).

From the figure above, it can be concluded that the further the distance of the roadside features from the road edge, the fewer chances for them being hit by vehicles going astray or moving off the road and less severity of accidents.

Most of the observed roadway infrastructure design related hazards are a result of uncoordinated design by the urban infrastructure operators in the road corridor. Some of the features in the design are copied from different standards. Procedures and methodology standards evolved from a wide range of empirical research conducted in those countries. However, procedures were developed for typical conditions found in developed countries where traffic is more homogeneous and regulated. Consequently, they cannot be applied successfully in traffic conditions that are significantly different such as those prevalent in Dar es Salaam, where there is a higher rate of disturbance to traffic flow from the operation of urban infrastructure. At times even the designs are often not fully implemented due to insufficient available funds.

Furthermore, investigation of the roadway identified the problem of sagging and open road surface without warning to users. It also ascertained existence of narrow and deteriorated shoulders in the road corridor. Then, there existed hazardous open drainages without provision of crash barrier and inadequate drainage systems.

Some of the ditches were wide and steep. There were no guide rails provided along ditches to prevent road users from falling in the ditches. The related hazards shown in Figure 5.20 are typical accidents occurring on road corridors due to design. The design of such drainage systems compromises safety of movement.

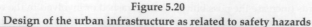

Figure 5.20
Design of the urban infrastructure as related to safety hazards

Source: Field data survey (2007).

5.6.2   Surface Condition-related Safety Hazards in the Road Corridors

(i)   Reinstated Road Section Excavated for Repair

Sewer deterioration undermines travel way structure leading to premature failure of travel way. The poor workmanship and inadequately reinstated road section excavated for repair are hazards to the road users.

Figure 5.21 gives a typical incidence of a road section cut for underground sewer pipes repair which was poorly reinstated. It resulted into safety hazards due to unevenness of the road surface and lack of prior warming to road users.

Figure 5.21
Poor reinstated road section and accident on the same section

Source: Field data survey (2007) **(ii) Problems with Water Infrastructure**

Problems with water infrastructure included leakage of water pipes under the roadway, water pipes cut by construction equipment or by other authorities when repairing other infrastructure networks, vibrations generated during road rehabilitation and excavations that break/deteriorate water mains.

Figure 5.22 illustrate leakages and cut water pipes, which led to the flooding in the road corridor. Consequences of such leakages are interruption to traffic and hazards to safety of movement.

Figure 5.22
Water layer on road surface due to water pipes defects

Source: Field data survey (2007)

(iii)   Unprotected Work Zones

Frequency of accidents in unprotected work zones was higher than on clear road sections. Observed risk of safety of movement on the work zones, especially during the night and in bad weather conditions, can be met with improved visual guidance and prior information to the public, which seemed to exist, especially when underground infrastructures were under repair on road corridors of the study area.

Figure 5.23
A vehicle accident in an unprotected excavation in study area

Source: Field data survey (2007)

It is important that work zones have adequate advanced warning. According to Ogden (2000), work zones can be hazardous to road users, when a work zone is left useful while work is not in progress. A number of road users such as pedestrians, cyclists, motorcyclist and vehicles have been falling in excavated pits. A typical unprotected work zone accident is illustrated in Figure 5.23.

iv)   Utility Cuts

Figure 5.24 shows an example of a pit typically found in the middle of most carriage way excavations to allow repair of underground sewer pipes. Excavated pits were left open on the carriage way for a long time, so they exposed pedestrians and motor cyclists to danger of falling in the pit and imposed safety hazards to the road users.

Excavation of road sections for underground infrastructure repair weakens the road layers which lead to deterioration of the network. This results to rutting, potholes and loosening of materials that could cause loss of steering control.

Figure 5.24
Excavations within roadway for repair of underground sewer pipes.

Source: Field data survey (2007)

(v)   Hazardous Manholes on Road Surface

Hazardous manholes were found to be potential safety problems on the road surface in road corridors, as they were found to be uncovered and protruded above the road surface. Uncovered manholes on the road surface either due to wear or vandalism were mostly reported as potential hazards to the safety of movement of people and vehicles in the road corridors as shown in Figure 5.25.

Figure 5.25
Uncovered manhole on road surface along Bagamoyo Road

Source: Field data survey (2007)

Moreover, most of the manholes placed on the carriageway on the road corridors under study area were found to have different levels with the road surface due to poor workmanship and compaction as illustrated in the Figure 5.26. These result in cracks and potholes on the road surface and are safety hazardous spots on the carriage way.

Figure 5.26
20 mm cover above carriageway in wheel path

Source: Field data survey (2007)

v)   **Leakages on the Road Surface**

Flooding is caused by leakage of underground pipes as shown in Figure 5.27. Faulty and operation interaction of underground infrastructure coupled with poor designs and poor maintenance of drainage system cause flooding on the road surface. Most of the drains have been insufficient due to the ever growing population in the city. Flooded roads prevent drivers from seeing what is on the road surface.

Figure 5.27
Flooding on road section at Chang'ombe along Mandela road

Figure 1 Source: Field data survey (2007)

Most of the road sections in the city are prone to water flooding due to leakages of either underground water pipes or sewer pipes. Water layers on the road surfaces are slippery and have been contributing factors to many accidents involving water layers on road surfaces.

5.6.3   Road Edge and Adjacent Land-Related Safety Hazards

(i)   Road Edge

The roadside area which is immediately outside the shoulder has a strong impact on the safety of movement. The recommended slope should not be less than 1:4. The desirable slope ranges from 1:6 to 1:7 in an area of 2-2.5 metres.

Infrastructure characteristics related to location of roadside features within the road corridor mostly determine involvement of features in the road traffic accidents. Seriousness of types of accident frequencies of vehicles leaving the road and hitting the roadside features is known to contribute between 18 and 42 percent of fatal accidents.

Table 5.13
Fatalities and injuries by most harmful events

Aggressive roadside object	Fatalities		Injuries	
	Frequency	Percent (%)	Frequency	Percent (%)
Vehicle to vehicle	17,495	40	1,721,000	51
Fixed object (poles)	9,239	21	503,000	15
Pedestrian/cyclist	7,481	17	114,000	3
Overturn	6,698	15	186,000	6
Other/unknown	2,192	7	839,000	25
TOTAL	43,825	100	3,363,000	100

Source: Tanzania Traffic Police Report (2007)

Indications of seriousness of the situation are illustrated in Table 5.13. Fixed objects on the roadside are harmful and account for seasonal greater percentage of total fatalities than they do for total injuries in the road corridors in the study area

(ii)   Adjacent Land Use Characteristics

Dar es Salaam is growing without adequate plan, which caters for existing mixed up land use patterns. The city's infrastructure network is not only obsolete but also inadequately planned to meet current demands. Adjacent land use along the road corridors in Dar es Salaam city can be categorised into three groups as listed below:

• **Residential only**
This includes residential houses and social institutions like schools, offices, hospitals and recreational grounds.

• **Commercial and residential**
This includes all residential features, shopping malls, hotels, pubs, possibly social halls.

• **Industrial**
This includes factories and warehouses.

The aforementioned categories can rarely be distinguished in most parts of the city. There is increasing violation of town planning, which leads to mushrooming of squatters, mixing up of commercial activities within residential areas including street vendors along road corridors. There are relationships between adjacent land use patterns and safety of movement of people together with vehicles in the public road corridor. Where there is a high interaction of people and vehicles, the areas have a higher exposure risk to road accidents. Collected accident data from Traffic Police (Traffic Police data base 2000 to 2007), indicated that accidents prone areas in the city are areas which have a high population interacting with vehicle movements as illustrated in Figure 5.28.

Figure 5.28
Vendors the walk way at Ubungo along Morogoro Road

Source: Field data survey (2007)

It is known that land use patterns can have various traffic safety and health impacts. High density clustered development patterns tend to increase traffic density (vehicle per lane-mile), which tend to increase crash rates per vehicle-mile within the area as observed by Kenworthy (1999).

Safety of movement in the study area was also noted to be influenced by the by-side friction as shown in Figure 5.29. Side friction is defined as a composite variable describing the degree of interaction between traffic flow and activities along the side(s) as well as sometimes across or within the travel way (Bang, *et al.* 1995). Activities disrupting traffic flow include blockage of the travel way (reduction of effective width) due to repair and rehabilitation work of the infrastructure system on the road corridors, public transport vehicles which may stop anywhere to pick up and drop passengers, pedestrians crossing or moving along the travel way, flooding on the road surface and trading activities.

Activities carried out in adjacent land use seriously affect traffic flow on road corridors resulting in reduced capacity of infrastructure and safety of movement. Therefore, it is important that the adjacent land use be properly managed to minimise costs and adverse impacts. If access is not controlled on busy roads, the function of the road can change, thus, reducing both its capacity and safety. A typical example is Mandela expressway where nearly all adjacent properly have direct access to the road. That has changed the road from a primary arterial road of high mobility to a lower category road (collector/distributor), whose primary function is access and has made it less safe.

Figure 5.29
Roadside friction along road corridor

Source: Field data survey (2008)

5.7 Relationships between Road Safety and Road Corridor Environment

5.7.1   Road Corridor Environment

Establishment of road corridors is primarily to provide a route from one place to another. But once established, roads encompass a range of activities. Apart from other activities, it is also a service corridor for electricity, drainage, water supply, sewage and communication networks. The width of the road corridors varies from one link to the other. Width variations of the links on the road corridors under the study are categorised according to the road hierarchy.

In Dar es Salaam city, the road corridors are characterised by uncoordinated urban infrastructure and queue due to failure of performance of infrastructure system in the city. The responsibility of managing the corridors in the city rests on the two government organizations, which are the Ministry of Infrastructure and Development, and the Prime Minister's Office, Regional Administration and Local Government. The deteriorated road corridor environment, reckless driving coupled with conflicts over the roadside use makes road corridors a complex area to manage.

The road corridor environment is affected by urban infrastructure repair and rehabilitation. The conditions in the city are worsened by leakages of underground pipes. Uncoordinated operations of underground infrastructure system and random location of underground infrastructure assets have also affected the movement and safety of movement of people including vehicles in the road corridor.

5.7.2   Safety of Road Corridors

Road safety is achieved through existence of a safe road corridor environment. It is constituted by mutual interaction of three elements, which are human, vehicle and infrastructure. These are primary factors in accidents. In order to ensure safety, there must be an existing environment that provides harmony between the human factor and the other two factors.

There is a relationship between road safety and road corridor environment, which is clearly illustrated by existence of incidences that involved mutual occurrence of accidents within a particular feature of the road corridor. The relationships were established by looking upon contribution of characteristics of road corridor environments at the area noted to be prone to accidents. Also contribution of urban infrastructure facilities into impairing safety of movement and incidences of accident occurrence associated with such facilities and the characteristics of adjacent road corridor activities and human behaviour and their contribution to intervention of fatalities / injuries and damages, was established.

Figure 5.30
Typical scenes of pedestrian and vehicular interaction in the corridor

Source: Field data survey (2007).

Figure 5.30 is an example of interaction of vehicles and pedestrians at Ubungo along Morogoro road. Pedestrians, like cyclists are vulnerable to traffic. Pedestrian death risk shows that even such a slow vehicle speed of 50 km/h kills 60 percent of pedestrians when hit (SWEROAD, 2004).

5.7.3   Road surface and safety of movement

Road surface conditions as said earlier were established to have a significant influence on safety of movement in road corridors in the city. Effects of road surface were observed and identified to be adversely affecting safety of movement in the study area.

Figure 5.31
Relationship between road surface conditions and accident rate

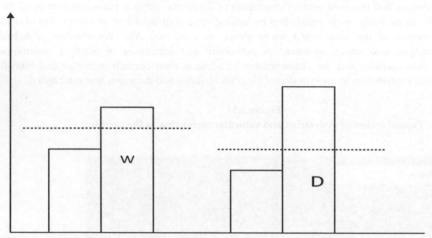

D – Represents dry
W – Represents wet
Source: VIT (1995).

Figure 5.31 represents an explanatory model of the relationship between accident rate and road surface conditions. During rains the accident rate is higher on worn rutted road surfaces than on less worn surfaces. During days of little rains or no rains the accident rate is lower on worn, rutted roads. The increase in accident rate with increasing precipitation is higher for roads with poor road surface conditions (VIT, 1995).

Although road surface contributes to road accidents, especially on poorly maintained roads, its effect is highly pronounced on vehicle operating costs. Maintenance work has a potential effect on safety (Transport Research Board, 1987).

5.8   Analysis of the Findings

The data were collected from target groups operating in the study area such as families living along the study routes, drivers operating in the routes, pedestrians walking along the routes and cyclists including other non-motorised traffic operating in the routes of the study area. Questionnaires were set to collect views from target groups of road users on safety of movement in the road corridor under study.

Thus, 500 questionnaires were distributed to targeted group road users in the following ratio: 250 questionnaires to families living along the five routes in the study area, 50 along the five routes in the study area, 100 to drivers operating along five routes, (20 to each route), 100 to pedestrians using the five routes in the study area (20 to each route), and 50 to cyclists and other non- motorised vehicles (10 to each route).

5.8.1   Respondents Profile

The study involved a total of 500 respondents, comprising 280 (56 percent) males and 220 (44 percent) females, located along five major roads in Dar es Salaam region, namely, Morogoro, Kilwa, Bagamoyo, Nyerere and Mandela. The distribution of respondents was according to their residence and road location. Respondents were randomly selected from various age groups including children, youths, adults and the elderly. Table 5.14 presents a summary of the respondents' ages.

Table 5.14
Distribution of respondents by age (N=500)

	n	%
Less than 18 years	18	3.6
18 years to 25 years	102	20.4
26 years to 59 years	371	74.2
60 years and above	9	1.8
Total	**500**	**100.0**

Source: SPSS result

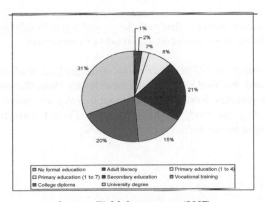

Source: Field data survey (2007)

The findings from respondents showed that road side infrastructure increases safety hazards in the road corridor in different proportions. Most of roadside infrastructure increases severity of road traffic accidents. Pedestrians and cyclists were reported to be in the high risks group.

Figure 5.33
Roadside infrastructure effects on safety of movement

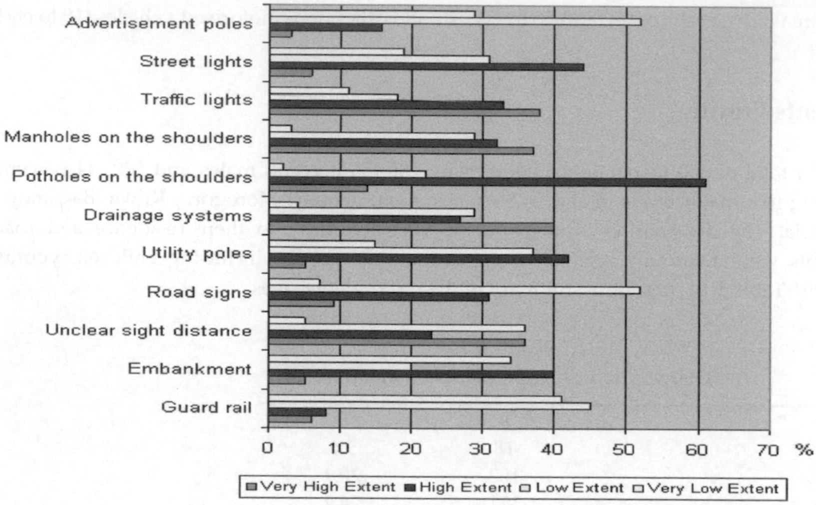

Source: Field data survey (2007)

Figure 5.33 shows the extent of effects of roadside infrastructure to safety of movement in Dar es Salaam road corridors. Survey showed that steep open drainages, traffic lights, open manholes and unclear sight distance had very serious effects on safety of movement and therefore increase severity of road traffic.

On the other hand, the survey depicted that potholes, street lights, utility poles, embankment, traffic lights, manholes and road signs have serious effects on safety of movement.

Figure 5.34 shows the extent to which roadway infrastructure in road corridor affects safety of movement. Survey indicated that road surface conditions mostly affect the safety of movement in road corridor, followed by the potholes, leakage of underground water and sewer pipes, sewage openings, manholes including cutting the road for crossing underground infrastructure. Other roadway infrastructure affects safety of movement in lower proportions.

Figure 5.34
Roadway infrastructure effects on safety of movement

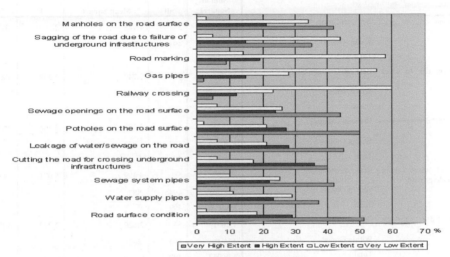

Source: Field data survey (2007)

5.8.2   Safety of Movement Element Ratings

Safety of movement was analysed on various corridors and means of transport. The analysis of 500 respondents involved in the survey was used to evaluate the extent to which various features on the road corridor affected safety of movement on each element of the road category.

Using the Pearson Chi-square, all ratings were seen to be significant at α=0.05. This implies that all responses are as shown on each frequency table. However, using the Analysis of Variance Test (ANOVA) most of the ratings presented in each element proved to be significantly different at 5 percent significance level. With the exception of gas pipes, road markings and road signing, the rest of the elements of safety movement along the corridor proved to be significant (as shown in Table 5.15). Chi-square proved the same in each element at the same level of confidence interval of 95 percent (Annex III).

Table 5.15
Analysis of variances of the element of safety movement

		Sum of Squares	df	Mean Square	F	Sig.
Unsafe movement due to road surface	Between Groups	9.915	5	1.983	2.762	.018
	Within Groups	354.635	494	.718		
	Total	364.550	499			
Unsafe movement due to water supply	Between Groups	13.702	5	2.740	2.559	.027
	Within Groups	529.050	494	1.071		
	Total	542.752	499			
Unsafe movement due to water supply pipes	Between Groups	16.633	5	3.327	3.108	.009
	Within Groups	528.719	494	1.070		
	Total	545.352	499			
Unsafe movement due to cutting road for cross underground	Between Groups	22.497	5	4.499	5.071	.000
	Within Groups	438.303	494	.887		
	Total	460.800	499			
Unsafe movement due to leakage of water/sewage on road	Between Groups	12.601	5	2.520	2.919	.013
	Within Groups	426.461	494	.863		
	Total	439.062	499			
Unsafe movement due to potholes on the road surface	Between Groups	9.444	5	1.889	2.555	.027
	Within Groups	365.234	494	.739		
	Total	374.678	499			
Unsafe movement due to sewage opening on the road surface	Between Groups	20.224	5	4.045	4.436	.001
	Within Groups	450.424	494	.912		
	Total	470.648	499			
Unsafe movement due to railway Crossing	Between Groups	16.740	5	3.348	4.229	.001
	Within Groups	391.138	494	.792		
	Total	407.878	499			
Unsafe movement due to gas pipes	Between Groups	6.575	5	1.315	2.028	.073
	Within Groups	320.313	494	.648		
	Total	326.888	499			
Unsafe movement due to road marking	Between Groups	1.404	5	.281	.443	.819
	Within Groups	313.274	494	.634		
	Total	314.678	499			
Unsafe movement due to sagging of the road	Between Groups	6.833	5	1.367	1.396	.224
	Within Groups	483.717	494	.979		
	Total	490.550	499			
Unsafe movement due to manholes on road surface	Between Groups	26.553	5	5.311	6.364	.000
	Within Groups	412.205	494	.834		
	Total	438.758	499			

Source: Field data survey (2008)

5.8.3 Average Safety of Movement

All the ratings being significant average safety of movement across all means of transport was 2.2 as indicated in Table 5.16 meaning that all respondents had an average agreement that mentioned elements led to unsafe movement to a "high extent". Moreover, the standard deviation is minimum (0.923), reflecting consistency in the feedback of the respondents.

Table 5.16
Average safety of movement ratings

N	Valid	500
	Missing	0
Mean		2.20
Std. Deviation		.923

Source: Field data survey (2008)

5.8.4 Factor Analysis of the Elements of the Safety of Movement

Factor analysis was applied in this study so as to reduce the number of elements that had to be discussed in detail that had greater influence on safety of movement. Moreover, Kaiser-Meyer-Olkin (KMO) and Bartlet's test confirmed that the data were fit to undergo factor analysis without limiting the number of factors. Kaiser-Meyer-Olkin Measure of Sampling Adequacy Measure varies between 0 and 1, and values closer to 1 are better. A value of .6 is a suggested minimum. Thus, for this KMO is 0.934. Bartlett's Test of Sphericity tested the null hypothesis that the correlation matrix is an identity matrix. An identity matrix is the matrix in which all diagonal elements are 1 and all off diagonal elements are 0 thus, rejecting the null hypothesis. Thus, with the case of this study the test is 0.000 which makes the data in this study significant. The results are presented in Table 5.17.

Table 5.17
KMO and Bartlett's Test

Kaiser-Meyer-Olkin Measure of Sampling Adequacy.		.934
Bartlett's Test of Sphericity	Approx. Chi-Square	10301.868
	df	66
	Sig.	.000

Source: Field data survey (2008)

Examining communalities, rotations and extractions of appendices, two factors were identified to have high load factors. Then, road surface and sewage systems had high load factor compared to other elements.

5.8.5 Analysis and Hypotheses Testing

In addressing the second objective, the relationship between infrastructure assets and average safety of movement is identified. To achieve this linear regression analysis, relationship between fault interactions and average safety of movement ratings and fault interactions have been developed. Moreover, the relationship between the ratings for the design and maintenance and average safety rating movements has been developed.

Hypothesis two

Null Hypothesis: There is no negative relationship between fault interactions and average safety of movement of link in a corridor.

Alternative Hypothesis: There is negative relationship between fault interactions and average safety of movement of link in a corridor.

By examining the model test of regression, the data deserved the test as shown in Table 5.18 as it is significant. It implies that the null hypothesis is rejected at 5 percent significance level and it can be concluded that there is negative relationship between fault interactions in a link and average safety of movement of the link in a road corridor.

Table 5.18
Analysis of variances (ANOVA)

Mode		Sum of Squares	df	Mean Square	F	Sig.
1	Regression	.052	5	.010	.011	0.000(a)
	Residual	286.698	318	.902		
	Total	286.750	323			

Source: Model results (2008)

Moreover, by examining the presented type of relationship, the following were deduced. R=13 percent, which shows a weak negative relationship between fault interactions and average safety of the link in a corridor. It means that as fault interactions increase, rating on average safety decreases reflecting that there is very high extent of being unsafe. However, the adjusted R^2=52 percent which reflects that 52 percent of average safety of movement is caused by fault interactions in a link.

Furthermore, observing at one coefficient (road aspect) and the constant of the relationship are as well observed to be significant. Thus, a linear relationship which is indicated in Figure 5.19 can be stated as average safety movement = -0.13 Fault interaction + 1.995

Table 5.19
Analysis of coefficients (a)

Model		Un-standardized Coefficients		Standardized Coefficients	t	Sig.
		B	Std. Error	Beta	B	Std. Error
1	(Constant)	1.995	.660		3.024	.003
	Action-Affected-Road	-.013	.064	.014	.210	.000
	Action-Affected-Water	-.007	.154	-.005	-.048	.962
	Action-Affected-Sewage	.001	.070	.002	.015	.988
	Action-Affected-Tellcom	.004	.337	.004	.012	.990
	Action-Affected-Power	-.002	.165	-.004	-.012	.990

Source: Model results (2008)

5.9 Conclusion

It was found out that the affected movement and safety on road corridors was a result of lack of coordination between the actors. Most incidences were due to lack of knowledge of underground infrastructure network, uncoordinated operations of the infrastructure, current practice of land use planning and poor workmanship.

Contributions of operational and faulty interactions among infrastructure assets were found to affect movement and safety. The FOI also reduced available shared space for usage by others when operational activities were fully undertaken by one or more urban infrastructure providers within the road corridors, resulting into obscuring the passages. It was also established that there was a tendency of having frequent and prolonged maintenance periods of urban infrastructure due to ageing and obsoleteness of network facilities that resulted into permanent impairment of the road corridor. They lead to inflicting physical damage to features of other infrastructure assets, impairing road corridor environment and interrupting services to the users.

Incumbent development and management approach affects movement and safety in road corridors. Undertaking of infrastructure assets development was uncoordinated among service providers, while development of networks was consumer-driven. Such uncoordinated development of urban infrastructure assets leads to friction among developers which eventually affects movement and safety.

In terms of management of the road corridor, it was found out that there were gaps in the set up and effectiveness in undertaking the responsibility of controlling as well as monitoring urban infrastructure development and operations. Furthermore, it was found out that there was little effectiveness in enforcing the law based on Acts governing infrastructure development within the road corridor by responsible authorities.

Results from the analysis provide evidence not to reject the hypothesis that the typical urban public road corridor in Dar es Salaam is unsafe, primarily, due to operational and faulty interactions among the infrastructure assets in the road corridor as well as the approach to the development and management of the assets.

PART THREE: INTEGRATING INFRASTRUCTURE SYSTEM IN THE ROAD CORRIDOR

In the previous chapters of this thesis it has been shown that the road corridors in the study area have low performance as far as infrastructure is concerned. This is illustrated by low level of safety of road users in the corridor, low structural capacity, high congestion, mobility disruption and economic inefficiency. It was also established that low performance results from the interdependencies among the existing urban infrastructure within the road corridor. Such situation is noted through incidences causing deterioration on infrastructure and impairing movement and safety. This calls for interventions that can align these urban infrastructure interactions, to attain harmony in road corridor operations.

The Integrated High Performance Infrastructure (IHPI) approach has been developed in countries that have faced similar problems. A number of cities have adopted a similar approach after facing constraints emanating from urban infrastructure interactions. In this part of the thesis we present the case for an integrated infrastructure system, a system of high performance infrastructure in an urban road corridor. It is the result of an integrated approach to the development and management of infrastructure systems in the urban road corridor. The system is conceptualised as having the potential of attaining unconstrained mobility and safety, being implementable and cost effective.

This part of the thesis is structured into three chapters in which chapter 6 deals with the proposed system of integrated urban infrastructure, chapter 7 deals with development of a model for analysis of performance of the proposed system and chapter 8 presents performance characteristics of the proposed system in Dar es Salaam and how to achieve its implementation in the road corridor.

6

6.0   INTEGRATED INFRASTRUCTURE SYSTEM

As pointed out in previous chapters, the rationale for effective utilisation of resources such as land and finance necessitates sharing of location without jeopardising service delivery. This ensures adequate movement including raising safety levels. It leads to inevitable interdependencies among urban infrastructure. Any approach of development and management of any element of urban infrastructure must be executed as an integral part of all types of infrastructure.

This chapter presents an integrated infrastructure and road corridor system as an appropriate approach to mitigate the mismatches, which bring about impaired movement including safety in the urban road corridor. The integrated infrastructure system is derived from an integrated approach to the development and management of infrastructure systems in an urban road corridor. As a way of showing that the system yields the best, it is termed "high performance infrastructure in an urban public road corridor". It is conceptualised to be implementable and enhancing movement and safety in urban road corridor.

6.1   The Integrated High Performance Infrastructure Approach

The Integrated High Performance Infrastructure (IHPI) is viewed as a system that yields output in terms of delivered services, while complying with movement and safety norms. It is characterised by the harmonic interactive existence of integrated urban infrastructures within the road corridor, bringing about high performance with minimum impact on others, thereby promoting synergy through the way they are developed, operated and managed. By presenting the characteristics of IHPI, elements which constitute this system of the integrated high performance infrastructure are considered.

The concept of IHPI is built on three fundamental considerations that characterise the urban infrastructure development and management approaches. The first concept is about attaining synergy among components of the urban infrastructure system. The second concept targets containment of conflicts that arise upon availing services to consumers and the third concept is about attainment of sustainable improvement of movement and safety.

6.1.1   Characteristics of the Integrated High Performance Infrastructure

Some concepts available in literature reflect the concept of IHPI. Nielsen (2005) considers this idea to have unique characteristics such as the best planning, designing, and management practices of the urban infrastructure network. Based on this concept, IHPI is assumed to be composed of integrated infrastructure as well as integrated approaches of developing and managing the urban infrastructure.

Moreover, the European Performance Infrastructure Guidelines (2005) express this concept as a combination or a system of facilities and measures that has two main properties: The first property is a set of proven best-integrated infrastructure planning, designing and management practices as well as good engineering practices. The second property is minimising negative impacts of geographical interdependencies and enhancing synergies as a result of effective coordination of construction and maintenance of different facilities. This design ensures easy access of utilities and reduction of conflicts between adjacent or overlapping components as well as timely preventative maintenance activities. It is obtained by having effective coordination of facilities designing, construction and operations.

The proposed integrated high performance infrastructure, according to Nielsen (2005) is characterised by:

- Constituent facilities within the road corridor that result from the best practice of planning, designing, building, maintenance and operation.
- A management approach which applies the best management practices.
- Employment of the best practices to coordinate investments so as to develop integrated designs for the facilities.
- Yielding optimum performance and minimisation of environmental impacts for each component facility.
- Being organised in a way that minimises conflicts, enhances safety among constituents as well as promotes synergies. This is concerned with scheduling and coordination of disparate activities in order to minimise impacts on citizens and businesses.
- Integrated designs.
- Achieving integrated infrastructure planning and implementation.
- Achievement of integrated asset management.

The proposed set of the best planning and management practices include scheduling and coordination of disparate activities so as to minimise impacts on citizens and businesses, integrated coordination of construction, maintenance as well as special events and integrated approach to planning and designing of infrastructure. In addition, what are considered as good municipal engineering practices should include coordination of joint use of road corridor by utilities, joint use of aerial utility poles by two or more utility types (e.g. power, telephone, telegraph, CATV) and common utility placement, namely, joint trenching (combining two or more utilities in a common trench). A good trench combination is placing of water and sewer lines in the same trench provided that sanitation hazards will not result.

6.1.2   Integrated High Performance Infrastructure Requirements

In order to meet the objective of having infrastructure that yields the best services to intended users, some criteria have to be met by the established infrastructure. The criteria are termed as IHPI requirements. High performance is achieved by incorporation of specific requirements in the establishment of the structure. The IHPI requirements are categorised into major areas, namely IHPI development requirements and IHPI operational as well as management requirements. These requirements are partially independent and partially interdependent. These requirements originate from different sources such as needs analysis, standards, and literature review.

(i)   IHPI Development Requirements
These are requirements that are concerned with upgrading of existing infrastructure or provision of a new infrastructure and physical attributes of the infrastructure network facilities. They are derived from the best practices of planning, designing and construction in order to attain the required level of IHPI during development of urban infrastructure network corridor. The requirements are elaborated below.

- **Satisfaction of service delivery to intended users**

One of the requirements is that individual urban infrastructure network should attain intended user service delivery standards in terms of movement and safety, user satisfaction and should be economical. High performance infrastructure should not compromise one type of service at the expense of another. Development of the urban infrastructure network should meet these service dimensions at all planning, designing and construction stages.

- **Effective utilisation of land**

IHPI requires planning of the urban infrastructure network that takes into consideration accommodation of all infrastructure networks in the corridor. This harmonisation is in terms of avoiding conflicts of urban infrastructure, and maintaining suitability of corridor layout without impairing efficiency of the facilities. Each urban infrastructure network considers the life cycle under characteristics of each urban infrastructure network planning, designing, and construction of the urban infrastructure network in the corridor.

- **Compliance with respective standards**

An integrated design of urban infrastructure network and placement of facilities are dictated not only by physical limit of the road corridor, but also by standards and rules. The aim is to ensure that the provided systems observe minimum burial depth, separation distances as well as safety and health rules requirements. Examples are water and sewer lines to be separated by a minimum distance and/or elevation difference or placed in impermeable materials. Gas lines are also limited by dynamics of their substance transmitted to locations away from certain other utilities where leakage may result in collection of potential hazards. It is also important to observe the standard colour of each underground infrastructure laid in the road corridor.

Design of the roadside infrastructure requires attaining movement and safety. Electric and telecommunication poles should be spaced and distanced from the roadway at the recommended safe distance, and they should be made of appropriate material. An effective distance to aggressive objects should also be taken into consideration in order to minimise severity of road traffic accidents.

The drainage system design for draining storm water from the road surface and catchment area along the network should be designed with consideration of safety and unimpaired mobility as well as avoiding flooding on the road network. Planning and designing of new IHPI network should incorporate development features and sustainable urban safety that provide multiple benefits such as enhancing movement and safety, reduction of localised urban flooding and harmful environmental impacts.

Preparations for a master plan for IHPI network should incorporate key components of all urban infrastructure networks. In terms of execution of development, the urban infrastructure network development should be extracted from the shared master plan. The design of the urban infrastructure network should put into consideration availability of other urban infrastructure networks in the road corridor. Implementation of the urban infrastructure network project should synchronise the urban infrastructure network projects during construction and provide properly designed utility crossing facilities so as to prevent cutting of the road network.

(ii)   IHPI Operational and Management Requirements

The integrated operations and management of urban infrastructure network in the road corridor that bring about high performance of the urban infrastructure network are constituted of the following elements:

• Harmonisation of infrastructure facilities operational tasks, so as to minimise inflicting deterioration on service and physical facilities.

• Presence of a centralised database of facilities layout including conditions, which is accessible to all operators.

• Presence of continuous liaison among operators that enhances thorough communication among operators when individual activities are carried out within the road corridor.

• Establishment of an integrated urban road corridor management system that manages utilisation of the entire road corridor by ensuring existence of monitoring and evaluation. This will result into enhancing coordination among infrastructure operators in the road corridor by exchanging information about future development plans as well as plans for ongoing management of urban infrastructure.

(iii)   Best Engineering Practices

The best engineering practices are other features associated with IHPI. They ensure minimisation of impact to service delivered when urban infrastructures are being developed or while carrying out operational activities. They happen regardless of streamlining disturbance of activities to services delivery level. The best engineering practices cannot completely be avoided within the shared corridor. The following approaches present the best engineering practices, which are associated with minimum effect to service delivery:

• **Alternative contracting methods**

The methods encourage contractors to complete works in a timely fashion which reduces costs, minimises overall disruption to the community during traffic flow, gives financial incentives for completion of works ahead of the contract schedule and imposes liquidated damage for each day or hour that the contractors go over the planned schedule.

• **Joint bidding**

This approach streamlines roadway reconstruction and reduces environmental, social and economic costs, by including private utility works and public works in a common contract that the city bids out. Joint bidding agrees to clear methods of payment prior to construction and requires sealed arbitration so that if additional work arises it is negotiated in a onetime arbitrated session without negatively impacting construction progress or stopping work.

• **Coordination of utility work with scheduled pavement construction and rehabilitation programmes**

This approach involves utilising additional contract requirements applicable to work performed in presence of privately owned utility facilities. It also employs utility cut moratoriums. It does not allow any excavation or utility cuts for a length of time – preferably a minimum of five years – after a road has been resurfaced, repaved or reconstructed. Use of grant waivers for utility under emergency circumstances, preserves pavement lifecycle, minimises disruption, reduces long-term costs, and enhances movement as well as safety.

- **Employment pavement degradation penalties**

It is suggested that penalties should be used to enact provisions that enable the road authority to recover costs from ongoing maintenance of utility cut repairs and penalties that develop a fee schedule based on unit cost of new material, excavation, site mobilisation, design, construction management, and contingencies. There is inclusion of a unit fee for social costs such as disruption to area businesses, traffic congestion, and impacts to public health. Grant waivers are possible if the utility company employs trenchless construction methods or agrees to coordinate its work with more roadway works to give incentives to private utilities, to upgrade infrastructure during road reconstruction.

- **Development of effective design guidelines for utility cut restoration**

This would include consideration for a range of traffic conditions and environments. Also it presents design alternatives whenever possible. It includes a complete as well as comprehensive analysis and provides advice based on available construction technology and local expertise.

- **Developing high performance infrastructure guidelines**

Guidelines provide a roadmap for incorporating the 'best management practices' (BMPs) in the urban road corridor infrastructure. The guidelines are intended for use by planners, designers, engineers, public officials, and anyone else involved in constructing, operating, or maintaining the road corridor. The high performance infrastructure guidelines allow for a practical, incremental approach to implement the best management practices in maintenance and construction works of urban infrastructure network in the road corridor.

6.1.3   Potential Benefits of Integrated High Performance Infrastructure

The integrated high performance infrastructure and the design of advanced common utility placement help to resolve basic issues such as movement and safety, coordination, quality of construction, economics, competition and high performance service delivery. The potential benefits of the integrated high performance infrastructure are explained below.

- **Enhanced movement and safety**

The number of conflicts is minimised and hence this will prevent interruption of traffic flow, thereby enhance movement and safety in the road corridors.

- **Enhanced infrastructure performance**

Fewer excavation accidents occur when utility location is clearly identified. The study indicated that common utility placement caused no increase in the number or severity of incidents. It reduced accidental digging to buried utility facilities and thus minimised traffic flow interruption. So it enhances performance of services in the road corridor.

- **Reduction on land and development costs**

Joint use of road corridor by utilities minimises the amount of land required for all utilities and leads to potential cost savings. Cost of joint utility trenches can be much lower than separate trenches, particularly for previous undeveloped sites.

- **Enhanced coordination in road corridor usage**

Presence of a sound communication channel and liaison among utility providers and road authority departments improve coordination in road corridor usage. The enhanced coordination brings about the following:

- Synchronisation of work and sharing expenses by different entities.
- Minimisation of pavement degradation, together with utility damages from utility cuts.
- Minimisation of environmental impact on construction.
- Reduced construction delays plus impacts on traffic movement.

The IHPI is based on provision of services that on the whole meet safety integrated design and economic requirements. Benefits from integrated development and management also include minimised conflicts created by the layout and intersections of urban infrastructure by having explicit provision for passages of all possible urban infrastructure network facilities as an integrated part of road structure. This is economical in terms of space utilisation since space in the urban area is very limited because the road corridor is already built up and land in the vicinity is in competitive demand with other investment opportunities; It also attains to a large extent safety of movement of people and vehicles in the road corridor thereby enhancing interoperability of the urban infrastructure network. Layout of the urban infrastructure network should allow interoperability so as to avoid conflicts in undertaking operational activities and increasing safety risks.

6.2   Implication of IHPI in the Road Corridor

An integrated high performance road corridor is land dedicated or purchased for use and for the benefit of the public as part of a transportation system or utility infrastructure. This typically includes roadway, shoulders, curbs, gutters, sidewalks, alleys, shared paths and bridges. In addition, there are also public and private utilities located underground, at road level or overhead in the public road corridor.

The goal of the integrated high performance road corridor is to enhance movement and safety, while maintaining mobility in the public road corridor by establishing policies and procedures to effectively manage a variety of activities that occur within the public road corridor, for example, road construction, utility work and special events. The integrated urban road corridor should have a management system which monitors and evaluates responsibilities of managing right of the way along the corridor.

6.2.1   Characteristics of an Integrated High Performance Road Corridor

The integrated high performance road corridor is conceptualised as a strip of land that provides and accommodates many infrastructure systems of different forms. These infrastructure systems are water, wastewater and storm water. The road corridor must be comprehensively managed through authorities with ability to plan, authorise, coordinate, analyse and communicate the use of the road corridor. It is a system that has the following features.

- Planning, budgeting and designing are within multi-departmental input.
- Construction and maintenance of the infrastructure components must be effectively coordinated.
- Infrastructure and other facilities are designed for easy access.
- There is a range of amenities and enhancements that have the potential to benefit safety, quality of life as well as balance the needs of pedestrians, cyclists, public transport users and drivers.
- Conflicts between adjacent or overlapping components are reduced.
- Non-destructive maintenance techniques are used and timely preventive maintenance is undertaken.
- It has minimum negative impacts on movement, safety and the environment.

The road corridor must be conducive to walking and have an optimal balance between needs for pedestrians, motorcyclists, bicyclists, bus users and utilities. The road corridor must be managed comprehensively. This requires new processes and tools that improve the ability to plan, authorise, coordinate, analyse and communicate, permitting use of the road corridor to improve mobility during the infrastructure repair and enhance movement and safety.

6.2.2 Components of Road Corridor with High Performance Infrastructure

This can be illustrated from the movement and safety perspective, with the following necessary components of an integrated corridor with high performance infrastructure:

(i)   **Application of Appropriate Best Management Practices.**
- Having optimal balances of needs of users such as pedestrians, motorcyclists, cyclists, bus passengers, and motor vehicles within the streetscape.
- Coordination of pavement management system with other citywide infrastructure planning and construction work processes to reduce disruption and gain more out of infrastructure investments.
- Having sound administrative, regulatory and financial mechanisms to reduce the frequency and impact of road corridor construction by private utilities.

(ii)   **Use of Appropriate Technical Strategies.**
- Alternative contracting methods.
- Coordination of utility works with scheduled pavement construction and rehabilitation programmes.
- Incentives to private utilities to upgrade infrastructure during road reconstruction.
- Effective design guidelines for utility cut restoration.
- Coordination of placement of public as well as private utilities using common trenching and utility ducts to minimise environmental impact of installation and maintenance.

(iii)   **Multifunctional Optimisation.**
- Using pervious pavement to reduce storm water runoff and peak demand on storm water management infrastructure while providing an adequate driving surface for vehicles.
- Utilising trenchless technology to repair water-main infrastructure while minimising trench cutting and subsequent pavement degradation.

(iv)   **An Integrated Design.**
- Designing a roadway with a diversely planted median which functions as both a traffic calming device and a storm water bio-retention area that improves pedestrian safety, minimises storm water runoff, dampens street noise and improves air quality.
- Designing an accessible utility corridor for sub-surface utilities within the roadway will allow for easy maintenance, minimisation of road corridor disruption, extended pavement lifecycle and reduced environmental impacts from repeated excavations and disposal of sub-base.
- Designing a road corridor with a reduced impervious pavement area which will, among other things, substantially help to improve air quality, increase pavement durability and calm traffic.

6.2.3 Performance Potential of Integrated Corridor with High Performance Infrastructure

If adequately designed, an integrated corridor with high performance infrastructure will have the following potentiality:

(i)   **Raise the infrastructure values.**
Infrastructure developments in the road corridor will not only cease to be the burden and cost drain but also add economic, social, safety and environmental value at lower costs as well as becoming a source of revenue.

(ii)   **Balanced need of road users.**
The system will be able to balance needs of pedestrians, bicyclists, mass transit users, and drivers, and will offer a range of amenities and enhancements which will improve safety and quality of life, because it will be conducive to walking and that will optimally balance the needs of pedestrians, motorcyclists, bicyclists, bus users, and motor vehicles.

(iii)   **The potential benefits of IHPI are as follows:**

- Minimisation of conflicts between pedestrians and vehicles, making the experience of walking safer and more enjoyable.
- Reduction in motor vehicle congestion, and air pollution.
- Space for vegetation and plan which reduces the urban heat island effect, improve air quality and local microclimate, as well as add scale and beauty.
- Restoration of urban streetscapes that were degraded by excessive automobile activities.
- There would not be adverse impact on motorised traffic or emergency vehicle service.

(iv)   **Attaining coordinated joint use of road corridor and joint development.**
- Encouragement of efficient use of precious road corridor space.
- Reduction in pavement degradation, improvement of infrastructure life cycles and maintenance of value for municipal infrastructure investments.
- Reduction in work stoppages including construction delays.
- Minimisation of traffic congestion and associated emissions.

(v)   **Attaining road corridor management system.**
- Coordination of construction work decreases in construction work and costs.
- Reduction of the impact of construction on the environment and community.
- Optimisation of effectiveness and value of public investments in infrastructure.
- Minimisation of unanticipated work, which could lead to pavement degradation.

(vi)   **Placement of public and private utilities using common trenching and utility ducts**
- Enables different entities to synchronise work and share expenses.
- Minimises pavement degradation from utility cuts.
- Minimises environmental impact of construction.
- Easy location and identification of utilities.
- Minimisation of incorrect excavations, utility damage and interference.
- Reduction in construction delays and impacts.
- Reduction in traffic congestion as well as vehicular emissions.

6.3   Conceptual Plan for Integrated High Performance Infrastructure

Based on the stipulated requirements of an integrated corridor with high performance infrastructure in a road corridor, the integrated corridor would be conceptualised with high performance infrastructure in the space within the road corridor that contains a complete street and integrated utilities' space. Such utilities' space is termed utility corridor, an area of road corridor designated for joint location of utilities.

It defines the pre-conceived ideals about an expected urban infrastructure design philosophy which leads to provision of specific functions layout, physical forms and basic structures. Designs of the layout of an urban infrastructure network should be in such a way that they support each other and not conflict with each other. These features are comprised of specific functions, layout, physical forms and the basic structures, the existence of which brings minimised conflicts and attainment of more synergy, while services are being delivered.

6.3.1   Complete Street Concept

A complete street is a street that enables safe and convenient access as well as travel for all users; pedestrians, cyclists, transit riders, people of all ages and abilities as well as freight and motor vehicle drivers. It is a situation that fosters a sense of place in the public realm. Complete streets improve mobility and urban livability by providing safe and comfortable transportation choices for people of all ages and abilities. They enhance the public realm with the incorporation of amenities such as vegetation, lighting, and other streetscape improvements. It also plays an integral role in addressing a range of issues that many cities are currently concerned with, including improving mobility, enhancing movement and safety, improving access and healthy communities, and maximising the use of scarce resources and funds. The various zones that comprise a typical complete street include the pedestrian amenity zone, parking zone, bicycle zone, and roadway zone (Figure 6.1).

Figure 6.1
Zones that comprise a typical Complete Street

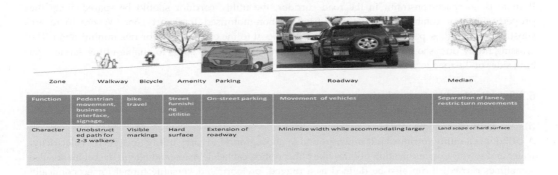

Zone	Walkway	Bicycle	Amenity	Parking	Roadway	Median
Function	Pedestrian movement, business interface, signage.	bike travel	Street furnishing utilitie	On-street parking	Movement of vehicles	Separation of lanes, restric turn movements
Character	Unobstructed path for 2-3 walkers	Visible markings	Hard surface	Extension of roadway	Minimize width while accommodating larger	Land scape or hard surface

Source: Field Data Survey (2007)

The sidewalk and amenity zones are the two areas that contribute to pedestrian environment. A safe, comfortable, and attractive pedestrian environment is vital for successful commercial districts and vibrant neighbourhoods. Pedestrian safety and comfort are related to the width of the sidewalk, the amount of buffering from traffic, illumination, and amount of pedestrian activity. An interface between building facades and sidewalk also contributes to pedestrian environment.

The amenity zone and sidewalk zone often complement each other and should be thought of as a system. Amenity zones help to buffer pedestrians from traffic, and may contain many of the amenity features that contribute to an attractive and vibrant streetscape; including street furniture, pedestrian lighting, street trees and street vegetation, loading/unloading room for on-street parking, kiosks, and public art, in constrained situations where the preferred sidewalk width is not achievable.

A complete street can also be a green street. The green street incorporates many green design features, including increased trees and greenery, infiltration landscaping to help absorb rain water and a unique storm water channel that runs through the length of the greenway. The flexible amenity is another unique feature of the greenway. It can accommodate parking, loading spaces and bus stops. The space can also be converted for pedestrian use during special events. The rolled curb allows car access while a textured pavement helps to separate the amenity zone from the road space. The goal of a green street is to enhance neighbourhood liveability while also providing ecological benefits through improved water and air quality, as well as potential habitat. The green street appears and functions differently from a conventional street due to its emphasis on incorporating Low Impact Development (LID) techniques and other green infrastructure such as street trees and vegetation. Opportunities for additional street trees and other vegetation are sought and/or created within a green street corridor in order to take advantage of the air and water quality benefits which they provide.

6.3.2   Zoning Concept

This is the concept of channelling road users within specific passages in order to minimise occurrence of conflicts that hinder movement and create high risks to safety and impair services. It is based on buffering of users, depending on level of vulnerability. Conflicts of service provider activities are higher when interacted with motorised vehicles. Utility zone placement is a way of minimising conflicts with transit lanes and it is an effective approach.

If there is no space constraint in the road corridor, the utility corridor should be spaced along the property line not combined with a motorised or non-motorised transport zone. Where there are constraints to utilities, passages have to be placed adjacent to the carriageway for risk minimisation. The crossing utility duct should be provided to each after short intervals so as to enable utility services on both sides of the road.

6.3.3   Utility Accommodation

(i)   Utility Corridors
A Utilidor is a utility corridor built underground or aboveground to carry utility lines such as electricity, water and sewer. Communication utilities like fibber optics, cable television and telephone cables are also sometimes carried. It can also be defined as a rugged, prefabricated versatile tunnel for economically housing underground utility services.

A utility corridor as defined by Webster's New Collegiate Dictionary is a passageway divided into compartments or routes. It can also be described as a conduit that can contain multiple utility systems. Such type of structure which is a comprehensive method of accommodating utilities has the potential to solve utility accommodation problems in urban areas with limited road corridors. It is a joint use facility or conduit constructed or installed within a road corridor that can accommodate a variety of utilities, either public or private, to minimise congestion of utilities within the road corridor and to facilitate co-location, maintenance, and access to utilities. Other terms of the utility corridor may include utility closet, duct bank and utility tunnel.

Figure 6.2
Structure to house the utility facilities for 12 companies

Source: Kuhn *et al.* (2003)

The utility corridor structures may be designed as shown in Figure 6.2. It is a large structure that provides a corridor as a walkway throughout the facility or as a smaller structure without a walkway and with accessibility provided at designated intervals by removal of the deck as shown in Figure 6.3.

Figure 6.3
Utility corridor with and without walkway respectively

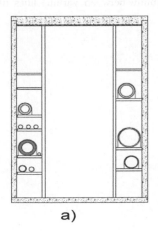

a)

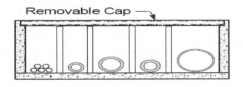

Source: Blaschke, *et al.* (2002)

The most comprehensive method of efficiently accommodating all or most utilities within the road corridor is with a utility corridor structure. With careful planning and designs, which meet all regulations governing utility placement, the structures can be used to accommodate not only telecommunications and compatible utilities but also traditional utilities such as water, wastewater, natural gas, and petroleum. In urban areas where the road corridor is extremely limited and space is congested with numerous utilities, the road authority may consider using utility corridor structures for accommodation of all or most utilities to relieve congestion and more effectively manage the available road corridor

Potential benefits for using utility corridor structures are to provide easy access and known location for all utilities in the road corridor. This reduces the likelihood of damage during subsequent road construction or maintenance projects. Installation of utilities within a structure requires coordination with other utilities, which helps reduce utility construction and installation time. It reduces the construction time, reduces the delay to the overall project and results in time and money savings for both road users and road authority. The structure can easily provide additional space for future utility utilisation and expansions without significant additional costs, as well as easy maintenance of underground utilities instead of cutting the road pavement surface. Potential benefits for using utility corridor structures are the following:

- Provides easy access and known location for all utilities in the road corridor. It reduces the likelihood of damage during subsequent road construction or maintenance projects.
- Enhancement of efficiency: Installation of utilities within a structure requires coordination with other utilities. It helps reduce utility construction and installation time. It reduces the delay to the overall project, resulting in time and money savings for both road users and road authority.
- Flexibility and economical for future expansion: The structure can easily provide additional space for future utility utilisation and expansion without significant additional costs.
- Cost effective in operation and maintenance: Easy connectivity and maintenance activities of underground utilities instead of cutting road pavement surface.

(ii)   Joint Trenching

Joint trenching is a method for addressing issues of compatibility and congestion. Figure 6.4 illustrates common trenching. This trenching design emphasises the need for appropriate spacing and placement of utilities. Utilised placement in this design ensures compatibility between various lines utilising the trench.

Figure 6.4
Joint trenching

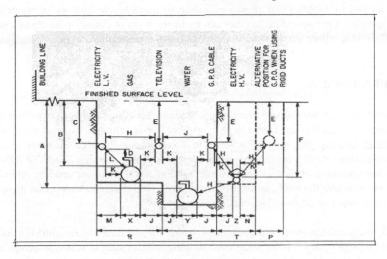

Source: Blaschke *et al.* (2002)

The original concept of common trenching is expanded in joint trench encased utilities. Joint trenching maximises utilities' installation activities, minimises disruption of existing utilities and simplifies locating the facilities for upgrades or repairs.

(iii)   Multiple Duct Conduits

Installing multiple duct (or multi-duct) conduits is another method for effectively managing utility installation within a limited urban road corridor. The multi-duct conduits are used instead of installation of numerous direct and buried individual cables/ pipes for individual utilities.

Multi-duct conduits can accommodate multiple utilities in one conduit, thereby requiring less space within the road corridor. Lines for individual utility companies are simply installed in separate compartments within one conduit, reducing overall costs for multiple installations. Multi-duct conduits can be a solution to maximise use of the road corridor without adverse impacts on the road corridor environment or motoring public in urbanised areas with limited road corridors. The conduit can be installed by using trenches, boring or horizontal directional drilling, which is beneficial in areas with bans against surface trenching for environmental reasons or road authority operations.

6.4   Integrated High Performance Infrastructure Alternatives

The integrated high performance infrastructure is a way of addressing diversity in existing layout of this study area whereby the respective road corridor cannot be best fitted with unique urban infrastructure layout. The IHPI alternative outlines take into consideration diversities and they are termed integrated infrastructure because they have to be integral structures of the road components within the urban road corridor.

Such alternatives are distinguished on the basis of utility accommodation alternatives derived from location and types of utility passages relative to other road features within the road corridor. Generally IHPI alternatives are formed by facilities that allow application of trenchless technology in repair and expansion of utility capacities. Corresponding facilities can be in the form of poles, single ducts, multi-ducts and utilidor. They may either exist as entities or combined overhead features, on surface or underground.

6.4.1   Hybrid IHPI Alternative

The hybrid IHPI alternative is an alternative having a utility zone that comprises both overhead and subsurface utility passage facilities. It is considered to be hybrid because it optimises the need of poles for accommodating overhead facilities like streetlights so as to accommodate passages of other utilities within the utility route. It brings about economy of space as well as materials through a reduced number of poles and by allowing utilisation of narrow space, while availing aside space for amenity and non-motorised traffic. It best fits where the road corridor is narrow like in old designed streets that preserve some cultural features and have limited space beyond the carriageway.

The electromagnetically transmissible services like electricity, telephone and data are transmitted through suspended cables held on large common poles while mass flowing utilities such as water, sewerage and gas are transmitted through the laid underground conduits. In this alternative utility the route is to be placed at the edges of the transit lane or park lane after positions of curbs, this serves as amenity for the vulnerable road users group. This alternative has the advantage of allowing flexibility in accommodating fast growing needs of ICT networks within the urban infrastructure without impairing mobility and safety for road users as well as incurring high cost of provision for underground passages.

6.5   Application of the IHPI Concept to Road Corridor in Dar es Salaam

Introduction of the integrated infrastructure system in Dar es Salaam will involve two fundamental issues to be established. The first step is to identify appropriate infrastructure features that can be adopted and second is the establishment of an institutional framework whose structure and obligations will support the set up of an integrated infrastructure system. The said institutional framework will have an operational package, which enables systematic, coordinated planning and programming of investments or expenditures, design, construction, maintenance, rehabilitation, renovation, operation, and in-service evaluation of physical facilities.

Application of the concept of the integrated corridor with high performance infrastructure is based on formulation of systematic interactivity roles of infrastructure agencies who are primary users of the corridor. It focuses on a system approach for infrastructure management that aims at having total asset management concept with all infrastructures operated within the corridor forming the constituent assets.

The proposed typologies on the study site are divided according to the type of road; it is 2 lanes 2 ways, or 4 lanes 2 ways, or 3 lanes 2 ways. These are types of roadway layout found in the study site. The proposals for location of underground and roadside infrastructure involve the integrated utility routes proposed to be located within the property line, where the road section transverses along the area, where land is available within the road corridor, and it is proposed to be located at the edge of the carriageway, where the road section transverses along the area within a limited road corridor.

6.5.1   Adoption of IHPI for Improvement of Utilities Infrastructure

The IHPI alternatives to be adopted for use within integrated corridor for the study area were chosen on the basis of following criteria:
- Available corridor space;
- Cost effectiveness of investment;
- Types of utilities to be accommodated; and
- Available appropriate technology to be used.

The typologies of the road sections of the study road corridors are divided into three groups as elaborated in the following sections;

(i)   Proposed Topologies for the 2 Lanes 2 Ways Undivided Roadway with Very Limited Roadside Land

Road sections that fall under this category include City Drive to National Social Security Fund (NSSF) along Morogoro Road, Ohio Street along Bagamoyo Road and Kivukoni to Bandari along Kilwa Road. These road sections are characterised by underground infrastructure on the road corridor, including water and sewer pipes and power and communication cables with deteriorated patch road surface due to underground repair. The sections also have manholes in the middle of carriageway and some are deteriorated. The roadside land use is a built up area with concentration of shopping centres and offices. Roadside land is very limited and very expensive to acquire. The proposed utility housing is utilidor with covered ditch sections for storm water runoff. The covered ditch area can be safely used by non-motorised transport.

Figure 6.5
Proposed topologies of 2 lanes 2 ways road sections

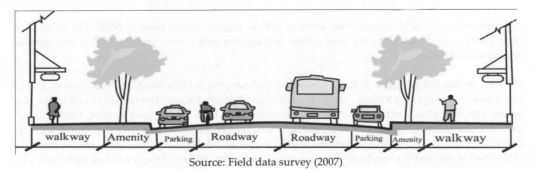

Source: Field data survey (2007)

Figure 6.5 shows the proposed topology of 2 lanes 2 ways road sections to be implemented along Morogoro road (i.e. from City Drive to NSSF Building), Bagamoyo road (Ohio Street) and Kilwa road (from Kivukoni to Bandari). The road sections have a limited road side land. The underground infrastructure under these road sections is proposed to be at the edge of the carriageway as shown in the Figure 6.6.

Figure 6.6

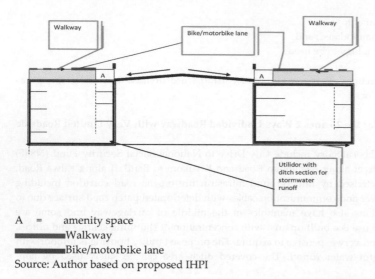

A   =   amenity space
████████Walkway
████████Bike/motorbike lane
Source: Author based on proposed IHPI

The utility routes adjacent to the carriage way are types of utility layout which minimise conflicts to operations and motorised traffic that enhance movement and safety. It also minimises interaction with the traffic flow. Such utility arrangement conflicts can occur between utility operations and non motorised traffic.

The road section along Morogoro Road, which is in this category, is City Drive to NSSF. The section is 2 lanes 2 ways. Land uses along the road section include shopping centres and the land is very limited along this road section.

The road section along Bagamoyo Road, which is in this category, is Ohio Street. The road section is from City Drive to Tanganyika Motors. The road section has 2 lanes 2 ways. The section is within the city centre with shopping centres and office buildings on both sides of the road section. Underground infrastructure under these sections is dilapidated, which calls for repair now and then in the sections through opening the road surface. There are also a number of manholes of which levels are not the same as the level of the road surface. Land is limited along the section and acquiring it involves high costs. The road section is frequently repaired because the underground utilities are also repaired frequently.

The road section along Kilwa Road, which is under this category is from Kivukoni to Bandari having 2 lanes 2 ways in the city centre. The section passes through the shopping centre with all underground infrastructures under the road corridor.

(ii)   Proposed Topologies for 4 Lanes 2 Ways Divided Roadway with Limited Land

This topology applies to road sections from NSSF to Dar es Salaam Fire Station along Morogoro road; from Tanganyika Motors to TAZARA along Nyerere road; Tanganyika Motors to Morocco along Bagamoyo road; and from Mwenge to Bandari along Mandela road.

Figure 6.7
Proposed typology of 4 lanes 2 ways plan

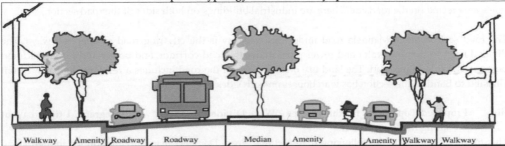

Source: Field data survey (2007)

Figure 6.7 shows the proposed typology of 4 lanes 2 ways with limited adjacent land for accommodating underground infrastructure. Figure 6.8 shows the proposed road network typology with utilities accommodation under the pedestrian walkway.

Figure 6.8
Integrated high performance road corridor with 4 lanes 2 ways

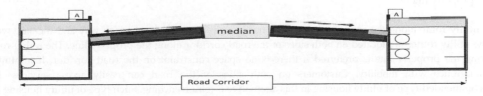

Source: Field data survey (2007)

Due to limitation of land in the road corridor, the underground infrastructure can be relocated from the carriageway to the edge of the carriageway as shown in Figure 6.8.

The road section along Morogoro road under this category is from NSSF to Dar es Salaam Fire Station. The section has 4 lanes 2 ways. Land use along the section includes office buildings and the land is limited along the section.

The road section along Bagamoyo road which is under this category is Alli Hassani Mwinyi. The road section is from Tanganyika Motors to Morocco. The section is 4 lanes 2 ways with water and sewer pipes under the road corridor while power and telecommunication cables are on the roadside. The section can accommodate utility housing on the roadside.

The road section along Kilwa road is from Bandari to Mtongani. The section is 4 lanes 2 ways with underground pipes under the road corridor. There are residential areas on both sides of the road and has the SabaSaba shopping centre, which is a big shopping centre in the area.

The road section along Nyerere road under this category is from Tanganyika Motors to Chang'ombe. The following section has only water and sewer pipes under the road corridor but power and communication poles are erected on the roadside. There are industrial buildings on both sides of the road section.

The road section along Mandela road under this category is the city ring road of 4 lanes 2 ways. The divided road section has water and sewer pipes under the road corridor, and power and communication poles on the side of the road. The land use on both sides of the road includes a residential area and from Uhasibu to Bandari the section has warehouses on both sides of the road.

(iii)   Proposed Topologies for the 4 Lanes 2 Ways Divided Roadway with Integrated Utility Layout along the Property Line

This topology applies to the following road sections: From Dar es Salaam Fire Station to Ubungo along Morogoro road, from Bandari to Uhasibu along Kilwa road, from Morocco to Mwenge along Bagamoyo road and from TAZARA to the Air port along Nyerere road. The road sections have space to accommodate utility housing along the property line.

- **Utility route along the property line**

Existing road corridors in the study sites are mostly housing underground utilities randomly, either at the median or at the walkway of the carriageway. The road corridor has four traffic lanes and a non-motorised lane on both sides utilising ditch sections. Regarding this type of layout arrangement, conflicts arise between utility operation and traffic flow due to unplanned settlements on both sides of the road. New construction or upgrading of the road corridor is being proposed with the utility route located along the property line.

A utility route along the property line allows traffic movement on the carriage way and the side walk. The utility routes are located on both sides of the road corridor along the property line. The utility route along the property line is preferred if there is no space constraint on the road corridor. It minimises interference with mobility. Customers on both sides of the road can easily access services. The recommended type of utility housing in this alternative is joint trenching. Such type of utility housing is a low cost method of installation, and it is not sophisticated. The surface can be treated, with little impacts on development materials.

The road section along Morogoro road, which is under this category, is from Fire Station to Ubungo. The section is 4 lanes 2 ways, divided road section. Land use along the section is residential with human activities along this section including local disorganised shopping centres and street vendors.

The road section along Bagamoyo road, which is under this category, is New Bagamoyo road; the section is from Morocco to Mwenge. The road section has water and sewer pipes under the road corridor and the power and communication poles very close to the road edge (1-2 metres). Land use along the section is residential with a few hotels and office premises.

6.5.2   Improvement of Location of Roadside Infrastructures along the Study Area

Tables 6.1 and 6.2 show recommended distance for relocation of the aggressive roadside infrastructure to the location desired for improvement of movement and safety.

Table 6.1
Distance from the edge of the carriageway

ADT	Speed km/h			
	50 and below	60	70-80	90 and above
< 1500	2m	3m	5m	6m
1500- 5000	3m	4m	6m	7m
> 5000	4m	5m	7m	8m

Source: Sondena (2007)

Table 6.2
Proposed distance of the aggressive infrastructure away from the road edge

Road	Road section	Traffic Volume	Operating Speed	Recommended Distance
Bagamoyo	City Drive-Movenpeak	>5000	30km/h	3 m
	Movenpeak-Morocco	>5000	50km/h	4m
	Morocco-Mwenge	1500 - 5000	60km/h	5m
Morogoro	Centre-Fire	>5000	30km/h	3 m
	Fire-Magomeni	>5000	50km/h	4m
	Magomeni-Ubungo	1500 - 5000	60km/h	5m
Nyerere	T/Motors-Uhuru	>5000	30km/h	3m
	Uhuru-Tazara	1500 - 5000	50km/h	4m
Kilwa	Kivukoni-Bandari	>5000	30km/h	3m
	Bandari-Uhasibu	1500 - 5000	50km/h	4m
Mandela	Mwenge-Ubungo	>5000	50km/h	4m
	Ubungo-Tazara	>5000	50km/h	4m
	Tazara-Bandari	1500 - 5000	60km/h	5m

Source: Author based on the Turner, (1990) recommendations
(Adapted from Turner, (1990))

6.5.3   Proposed Institutional Arrangements

In order to achieve the required IHPI, an alternative management system needs to be established from an effective management structure comprised of coordinating units for the infrastructure operators, in the road corridor as illustrated in Figure 6.9. Aiming at centralisation of authority for utilisation of the corridor, the entity is going to control utilisation through coordinating, planning, design, construction, maintenance, rehabilitation, renovation, operation, and in-service evaluation and management of information database.

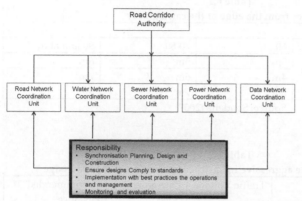

Source: Field data survey (2007)

In order to attain the best yields from the existing set up, the established coordinating unit in each urban infrastructure department will prevent overlapping of responsibilities and enhance efficiency in performance and prevent adverse effects of urban infrastructure on network development, operations as well as management of the road corridor. The coordination unit will ensure that operational aspects and management of urban infrastructure network are not inflicting deterioration on each other. Activities to be implemented in the road corridor are to be communicated to operators. That will enhance coordination of urban infrastructure network operations in the public road corridors, and sharing of information among the operators. The system should be developed to ensure that there is a liaison of activities in the road corridor, with other urban infrastructure operators. The infrastructure installation and operations guide is to be used to achieve improved movement and safety. The guide is attached as Annex III.

In terms of information management, there should be an Information Management System (IMS), which will centralise database in each coordination unit for all possible urban infrastructure facilities. That allows for proper documentation and sharing of information among operators. The proper documentation of urban infrastructure network and sharing of the data, especially proper location of where urban infrastructure network is located will enhance movement and safety as well as reduce operation and maintenance cost. The proposed management structure will enhance the following:

- Successful implementation of IHPI;
- Integrated urban infrastructure passage facilities;
- Synchronisation of standards;
- Coordination of corridors utilisation;
- Creation and sharing of information database;
- Best municipal engineering practices;
- Monitoring and evaluation;
- Effective infrastructure network database to ease Integration of UI network layout.

6.6 Conclusion

This chapter has dealt with measures for mitigating the existing impaired mobility and unsafe condition within urban road corridors as revealed in the previous chapters. The preceding sections show that the main factors that bring about impaired movement and safety are emanating from existing urban infrastructure incumbent deficiencies in urban infrastructure network development and management approaches. There is also a call for consideration of appropriate measures to address these deficiencies. This chapter has also pointed out fundamental concepts in terms of development and management approaches, which have to form mitigation alternatives that would be the basis for functional designs of an integrated infrastructure system. It has further centred into elaborating its characteristics and their potential benefits into attaining improved movement and safety within the urban road corridor. Effectiveness of this integrated infrastructure system as conceptualised in this chapter is subjected to assessment in the chapter 8, so as to determine its suitability should it be adopted in the study area.

7.0   ANALYSIS OF PERFORMANCE OF THE INFRASTRUCTURE IN DAR ES SALAAM

7.1   Introduction

This chapter presents a multi-variable decision support model for analysing road corridor features. All features of Road Corridor Safety Analysis (RCSA) were extended for this research. The RCSA model is used by engineers and decision-makers to support complex decision making. It is designed to evaluate safety of movement in the road corridor by assessing conditions of roadways and other assets in the road corridor as well as the design and location characteristics of the infrastructure in the road corridor. It can be used by engineers and managers for safety evaluation of existing and proposed infrastructure. Furthermore, the RCSA can help them to anticipate the potential vulnerability and proneness of a road segment to accidents and thereby help to mitigate and prevent potential risks.

This computer based model is specifically intended to help decision makers and engineers to identify and document management priorities; define factors that influence such management priorities; rank relative importance of these contributing factors as they relate to different management options; identify as well as assess knowledge including data information gaps; and prioritise research and management options.

7.2   Characteristics of RSCA Model

Development and operation of the RSCA model are based on conceptualisation of major aspects and interactions in an urban road corridor way such as travel, land use, road traffic and infrastructure characteristics as illustrated in Figure 7.1. These characteristic features are key variables in decision making.

Figure 7.1
Major aspects and structure interactions in the road corridor

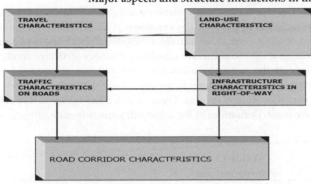

Source: Field data survey (2007)

The elements that constitute model variables are the conditions and interactions that affect safety of movement. Thus, the main model variables are as follows:

Road corridor environment: Road corridor conditions of underground, roadside infrastructure and roadway condition on each link and junction in the road corridors.

Underground infrastructure: This includes water supply network, sewerage (system) network, and electric power and telecommunication cables placed under each road corridor.

Open drains: A long narrow excavation designed or intended to collect and drain off surface water.

Seasonal variation in terms of rainfall: Weather variations reflecting rainfall regime in terms of average low and higher annual rainfall pattern.

Road surface: The road structure above the formation design to spread loading over the base and sub base.

Traffic factors: Traffic factors include traffic flow of vehicles and pedestrian and traffic speed.

Pavement design: Specification of structural elements of a road.

Geometric design: Specification of geometric elements of a road such as lane width, mean, shoulders, sight distance and cross-sectional element.

Traffic control factors: Traffic control facilities to control traffic in a road corridor.

Safety of movement: The traffic movement in a road corridor with no danger of being killed or injured in the process.

The state of each model variable is defined in terms of relative scales, which relate to the existing conditions and available empirical data as follows:

State category of urban infrastructure: Urban infrastructure variables evaluated for the model included road surface condition directly affecting safety of movement. The variable was considered in the model in order to evaluate its effect on the safety of movement. Other variables considered under the road surface conditions were pavement design, underground infrastructure condition, drainage structure condition, maintenance techniques and operations of the urban infrastructure network that affected safety of movement through affecting road surface conditions. Another factor considered under this category was season variations conditions through variation of rainy seasons. These factors affect urban infrastructure, corridor environment and safety of movement. Definitions of the urban infrastructure state categories are shown in Table 7.1.

Table 7.1
State definitions

Variable name	State category	Definition of the state
Road surface condition	Good	No visible defect, that is, the running surface is the 'as built' condition IRI<2.5
	Fair	Frequency of defects with medium severity. Frequent potholes and/or patches with extensive visible cracks IRI<5
	Poor	High frequency of defect with high severity. Extensive potholes and patches IRI<7
Pavement design	Good	Design with the required standards such as adequate pavement thickness and standard bitumen mix
	Fair	Missing one of the standard requirements
	Poor	Does not meet the required standards
Underground infrastructure	Good	Good standard for the underground infrastructure with no defects for example pipe leakages
	Fair	Fair stands for underground infrastructure with minimum defects observed
	Poor	Poor stands for underground infrastructure with extensive defects observed for example the leakages and burst of water and sewer pipes in the road corridors
Drainage infrastructure	Good	Shape and level of drains adequate. Drains functioning effectively, although minor silting may be evident
	Fair	Defects or silting evident and drainage capacity impaired
	Poor	Serious scouring or complete blockage affecting a significant length of sub-link including structure unsafe design
Seasonal variation	100 -250	Short rains season is described when the rainfall recorded was between 109.4mm to 573.2mm and it was recorded in the months of September, October, November and December
	250 – 400	
	400 – 550	
	250 – 450	Long rains season is described when the rainfall recorded was between 273.3mm to 796.3mm and it was recorded in the months of March, April and May
	450 – 650	
	650 – 850	

Source: TANROADS maintenance manual (2007)

State categories of geometric design: Geometric design of a road, which is the specification of geometric elements of the road, was considered in the model in terms of the number of lanes, lane width, shoulder width, median, sight distance and cross- sectional elements. These factors were evaluated to identify their effects on safety of movement in the road corridor. The states of these variables are defined in Table 7.2.

Table 7.2
State definitions of the geometric design parameters

Variable name	State category	Definition of the state
Number of lanes	2 lanes 2 ways, 4 lanes 2 ways	The numbers of lanes are defined to be 2 lanes 2 ways; this means it is an undivided carriageway with traffic flow in each direction through one lane and when it is 4 lanes 2 ways, it means there are 2 lanes in each direction. In this case it is called divided carriageway
Lane width	3m, 3.5 and 3.75	The width of a carriageway required to accommodate one line of traffic. 3m is the minimum lane width of the single lane; with 3.5m the lane is relatively wide, while 3.75 the lane width is wide
Shoulder width	None, 0-1m, 1-2m, 2+m	Paved or unpaved part of the road next to the outer edge of the pavement. The shoulder provides side support to the pavement and allows vehicles to stop or pass in an emergency. It is considered none when the shoulder is not there but varies from 0 to 2+m
Median	None, Present	Median is placed on the middle of the carriageway to separate traffic. None means there is no median; this type is also called undivided road section. "Present" means that the median is there and thus a divided road section.
Sight distance	0-25m, 25-50m, 50-100m, 100+m	Sight distance is the zone adjacent to the road which enables the driver to see what is happening on his/her sight and varies between 0 to 100+m. When it is 0 to 25 m it is termed "minimum" and vice versa
Cross sectional element	Open, embankment, cutting, Hill-side	Design "cross sectional element" is an open embankment, cutting and hill side when side slope represent on the road section have the above mentioned condition

Source: TANROADS maintenance manual (2007)

State Categories of Traffic Factors and Road Furniture: The traffic factors and road furniture variables are also included in the model to evaluate their effect to the safety of movement. Considered traffic factors are traffic flow and traffic speed. Table 7.3 presents the state definitions of the Traffic Factors and Road Furniture.

Table 7.3
State definitions of traffic factors and road furniture

Variable name	State category	Definition of the state
Traffic flow	<1000, <2500, <5000, >5000	The number of traffic (vehicles) moving on the road corridor in an hour is categorised to be low when it is less than 1000 and high when its more than 5000
Traffic speed	High, Moderate, Low	Traffic speed of a vehicle is stated to be high when the vehicle is moving very fast, moderate when the vehicle are moving at an average speed and low when the vehicle is moving slowly
Road sign	Good, Fair, Poor and None	A road sign is information posted on the roadside to communicate with the road users. It is good when the road sign is clear, visible and the massage is efficiently delivered. It is poor when it is damaged, the massage is faint and none when there is no road sign.
Road marking	Good, Fair, Poor and None	The demarcation on the road surface showing the centre line, demarcating shoulder and carriageway are stated good when they are clear and the massage is clearly sent to the user; they are fair when the demarcations are not clearly seen; they are poor when they are poorly placed; and none if there is to marking on the road section
Traffic light	Good, Fair and Poor	Traffic lights are electric signals used to direct the traffic. They are good when they are operating at full capacity; fair when are having defects regularly; and poor when they are not in operation

Source: TANROADS maintenance manual (2007)

Figure 7.2 shows the overall concept of the RCSA model. The model incorporates methods to identify and analyse potential problems as well as evaluate safety risks associated with the potential problems.

Figure 7.2
RCSA model

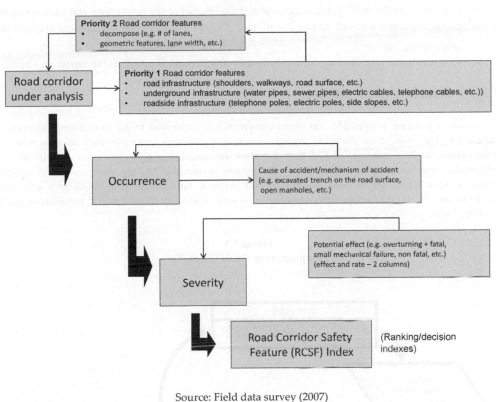

Source: Field data survey (2007)

It also assesses risks associated with features identified during the analysis and prioritises corrective actions. The risk is defined in terms of an index known as a Road Corridor Safety Feature (RCSF) index, ς which is mathematically defined as follows:

$$\varsigma = \sum_{0}^{i} \sum_{0}^{j} \sum_{0}^{k} \pi_i \chi_j \omega_k$$

(7.1)

Where; x_j and ω_k are occurrences and severity of accidents respectively. The higher the value of ς, the higher the accident risk in a road corridor. Thus, since the objective should be to minimise the ς, the objective function (Obj.) for engineers and decision-makers can therefore be defined as follows:

$$Obj. = Min \quad \varsigma(\pi, \chi, \omega)$$

(7.2)

Using a RCSF index ς to assess safety risk on a road corridor requires:

- Presence of a road corridor feature, π_i ;
- Rating of the severity ω_k of each effect of failure;

- Rating the likelihood of occurrence (x_j) of each cause of accident (i.e. for each mechanism of each accident);
- Calculation of the RCSF index, ς by obtaining the product of the two ratings and presence of the corridor features for each factor: $\varsigma = \pi_i x_j \omega_{k}$, where x_j and ω_k are occurrences and severity respectively, and
- Comparing safety of road corridor features and prioritisation of problems for corrective action.

7.3   Structure of the Model

As computer software is available, the major components incorporated in the model are illustrated in Figure 7.3. The three main elements, namely a set of **algorithms** to enable data investigation, an **information retrieval system,** and a **user interface** for providing easy access to algorithms and information retrieval systems were utilised. The system is intended to help decision-makers compile useful information from raw data, documents and personal knowledge so as to identify and solve problems and make decisions to reduce frequency as well as severity of road traffic accidents in public road corridors.

Figure 7.3
Components of the RCSA model

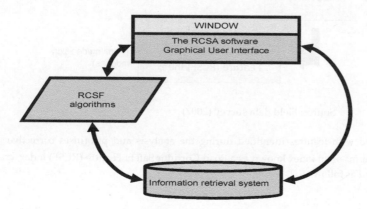

Source: Field data survey (2007)

The model is organised in a modular fashion and integrated into one system as shown in Figure 7.4. The main functional modules are data input module, data processing and analysis, presentation and knowledge exploration modules. Each of the modules consists of the main functions, sub-functions of various levels, and elementary functions. An elementary function practically equals menu commands. An information retrieval system was also incorporated.

The user interface provides links to functional modules and keeps information available whenever requested. The main goal of data input module is to facilitate easy data input. It provides models to support input, editing and visualisation of changes made.

Figure 7.4
Information flow during RCSA

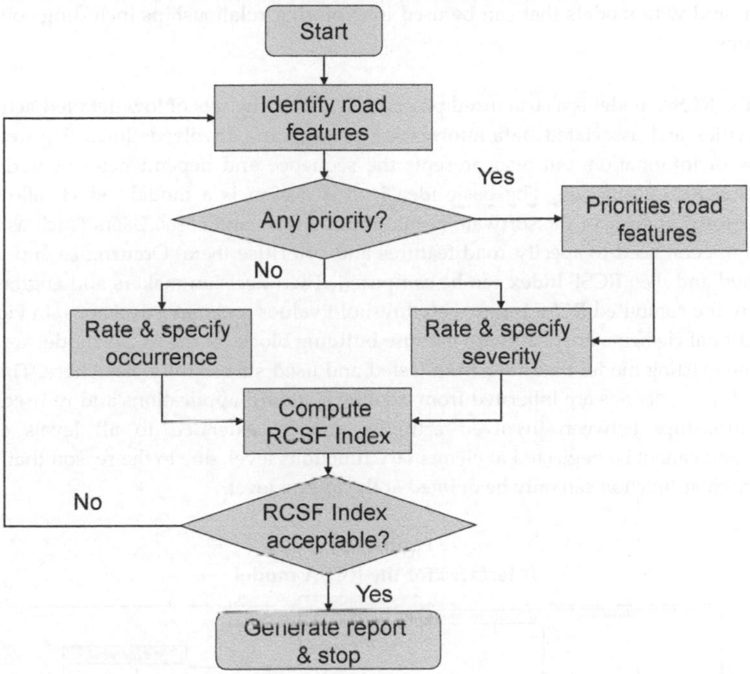

```
                        ┌─────────────┐
                        │    Start    │
                        └──────┬──────┘
                               │
                        ┌──────▼──────┐
            ┌──────────►│ Identify road│
            │           │   features   │
            │           └──────┬───────┘
            │                  │
            │                  │              Yes
            │            ◇ Any priority? ◇──────────►┌──────────────┐
            │                  │                     │ Priorities road│
            │                No│                     │   features     │
            │                  │                     └───────┬────────┘
            │         ┌────────▼──────┐         ┌────────────▼───┐
            │         │ Rate & specify │         │ Rate & specify │
            │         │   occurrence   │         │    severity    │
            │         └────────┬───────┘         └────────┬───────┘
            │                  │      ┌──────────────┐    │
            │                  └─────►│   Compute    │◄───┘
            │                         │  RCSF Index  │
            │                         └──────┬───────┘
            │      No                        │
            └──────────────◇ RCSF Index ◇────┘
                            acceptable?
                               │
                             Yes│
                        ┌───────▼────────┐
                        │ Generate report │
                        │     & stop      │
                        └─────────────────┘
```

Source: Field data survey (2007)

The main function of knowledge exploration utility is to support exploration and management of knowledge and provide users (such as decision-makers and engineers) involved in road safety analysis with effective access to information, regardless of their geographical location. It offers models for broad corridor processing of information. Knowledge exploration is well suited to computer assistance because the computer can effectively handle large amounts of information and avail such information in various different forms and formats. The motivation behind building knowledge exploration utility is to enable and facilitate reuse of existing knowledge. The goal is to enrich users (such as decision-makers and engineers) with key information needed for decision-making. The idea is to help decision-makers and engineers to familiarise themselves with previous decisions when analysing and judging a possible way forward for the road in question. The knowledge exploration module also accommodates models for supporting decision-makers to retrieve related information based on one or more search criteria.

The presentation module supports users to visualise processed data in various different ways (such as in graphical form and in tabular form) and has built in utilities to support sharing information, for example by sending processed electronic data to other stakeholders. It also comprises models, which support subjects in sending back complete analysis reports.

The information analysis module supports analysis of road safety features, (that is, field data). Furthermore, it comprises algorithms that can be applied in determination of priorities to give to various road safety factors, and in ranking road corridors according to likelihood including potential severity of accident. In addition, it also equips decision-makers with visualisation models (that is, for visualising processed data) and with models that can be used in exploring relationships including consequences of adopted solutions.

Theoretically, the RCSA model is a structured procedure, formed by sets of low detailed activities. Figure 8.5 shows activities and associated data/information process and involved flows. Figure 8.5 not only represents flow of information, but also presents the sequence and dependencies of various activities supported by the RCSA software. The basic idea in this design is a model, which allows for reuse, extension or removal of parts of the software at minimum efforts and costs. Users (such as the decision-makers and engineers) need to specify road features and prioritise them. Occurrence and severity rates are then specified and then RCSF Index can be computed. Then decision-makers and engineers therefore need to compare the computed RCSF Index with threshold values and make decisions. In Figure 7.5, only high-level functional elements are dealt with because building blocks of the RCSA model are mostly units or utilities of the existing model that have been tested and used successfully elsewhere. This means that embedded low-level functions are inherited from existing standard applications and re-used. Description of logical relationships between involved activities can be extended to all levels of functional decomposition and cannot be neglected at elementary functions level, due to the reason that input/output related to a particular function can only be defined at the lowest level.

Figure 7.5
Interfaces for the RCSA model

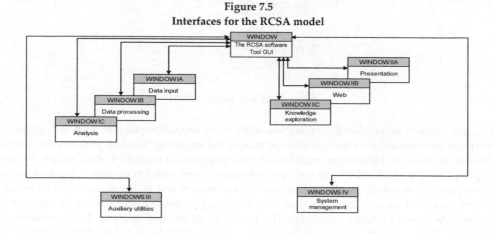

Source: Field data survey (2007)

7.4 Development of a Prototype

The component re-uses software development methodology, which according to Jones (1990), was adopted and used as a concept model for shaping as well as directing activities during development of the RCSA software. The main reason for choosing the components re-use methodology was efficiency considerations. It provided an economical way of creating new software in that implementation process largely involved identification and integration of existing codes (that have previously been tested and used in other areas and for other purposes) or fully-fledged functions, utilities or applications, sometimes without any sizeable modification; and using an interface as glue. The main challenge was how to interface components in an effective way in order to satisfy the requirements for the RCSA software, and how to synchronise the needed modification to existing codes. The strategy was to utilise components of widely used commercial software products as much as possible in order to speed up the development. Also it ensured portability and compatibility. As a large portion of the RCSA model was built using Microsoft technology, and source codes were written mainly in Visual Basic, Visual Basic for applications, Visual C++ and HTML. In general terms, there was no significant advantage from research point of view of developing everything from scratch since similar utilities were readily available in other existing software packages, and could be adopted as well as used to sufficiently demonstrate how computers can support the RCSA process.

Figure 7.6 shows a schematic representation of interfaces implemented to provide interactions between user and the RCSA software. In general terms, control over processing is exercised by the user (such as the decision-maker/engineer), who issues instructions to the system from the menu. Menus were preferred for the reason that they have an advantage of being quite natural and give the user strong control over the course of decision-making. However, they have a disadvantage of being slow when used in large and complex situations.

Figure 7.6
RCSA model

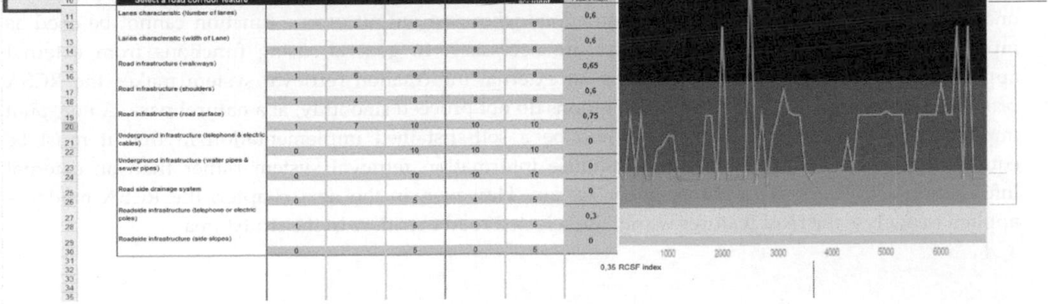

The RCSA model required more than one window. Therefore, its user interface was implemented as a multiple-document interface (MDI), which is the container for all forms. In other words, the MDI provides a parent window from which various functions of the RCSA model can be accessed and run. Application of the MDI means that the user of the RCSA model can open many windows and applications simultaneously. It brings users (such as the decision-makers/engineers) needed flexibility and convenience in accomplishing the RCSA activities. For example, during the analysis and interpretation of results, decision-makers and engineers can simultaneously have, in separate windows on a computer screen, information required when taking decisions.

7.5   Performance Evaluation

After developing the RCSA model, the pertinent issue was: "Can the model adequately do the job it is designed to do?" In addition, addressing this issue properly required providing answers to some questions, which included: Are theories and assumptions underlying the model correct? Is the model representation "reasonable" for the intended purpose of the model? Does the model's output behavior have sufficient accuracy for the model's intended purpose over the domain of the model's intended applicability? This section addresses these questions. It describes the process of verification and validation of the model system as well as details of validation in terms of characteristics and results from studies conducted to assess its suitability as a model for producing accurate or reliable results.

7.6   Conclusions

The RCSA model is important for two reasons. Firstly, there was a need for a model that would (i) assist in evaluation of infrastructure engineering and management strategies, that is to produce information that would allow professionals identify weaknesses in design and provide the basis for identifying the optimal designs; (ii) decrease time and costs for safety of movement evaluation and design, and (iii) provide better relative evaluation of urban infrastructure development and operational improvements, thus assist professionals and decision-makers in producing sound designs and evaluation of facilities for safety purposes. Secondly, the state of the art review revealed that there was no such a model that could be used for analysis of urban infrastructure development and management actions in urban areas of developing countries until now. The analysis of the characteristics of current major models showed that none of the existing models is suitable for analysis of development, operations and management actions on UI network in the road corridor.

A pilot implementation of the RCSA software was based on knowledge and concepts presented in previous chapters. The model can be used in real-world situations. Limitations of the RCSA model are mostly due to the fact that it was implemented based on existing applications and used external information and an information retrieval management system. The fact that, in some cases, the output(s) of a function may not be transferable to other functions (e.g. because of differences in data structures) is one of the viewpoints of this model. In situations where an output of one function cannot be used as input in another function, manual entries are necessary. In general, using functions from external applications to perform activities or using an external information retrieval system makes the RCSA process more sluggish, and sometimes operations do not proceed smoothly, at a natural pace. A non-pilot implementation of the RCSA model should be a self-sustained implementation in that it must be equipped with its own functions and in-house information retrieval system rather than an external information retrieval system management system. However, in the next chapter, the RCSA model is applied to analyse the road features within the selected road corridors in the study area.

8

8.0   SIMULATED PERFORMANCE CHARACTERISTICS OF THE PROPOSED SYSTEM

8.1   Introduction

This chapter presents results from a simulated performance of proposed integrated high performance infrastructure in Dar es Salaam. Simulation was done by inclusion of IHPI features as inputs into the RCSA; specifically it presents applications of the Road Corridor Safety Analysis (RCSA) model in evaluating the level of safety and effectiveness of IHPI on improving movement as well as safety in urban road corridors. The evaluation delineated movement and safety benefits in the road corridors that are attainable through implementation of IHPI.

8.2   Application of the RCSA Model to the Road Corridors

8.2.1   Validation of the Model

To validate the RCSA decision support model, case study areas along Morogoro road and Bagamoyo road were identified. These areas were the focus of data collection that could be used to evaluate the condition of the road corridor. The RCSA decision support model was considered adequate as long as data obtained from the study areas were statistically close to results from the model. Where substantial differences were observed, adjustments of weights of the model parameters were carried out until a concessional value was achieved.

After validation, the RCSA decision support model was used to assess effects of interactions of infrastructures on the level of safety of movement. Results from the simulation led into understanding the most critical urban infrastructure variables necessary for desirable safety performance of the road corridor. Furthermore, the simulation revealed likely effects of alternative infrastructure assets design, location and condition on safety in the road corridor; also it revealed expected level of movement and safety within a given section of an urban road corridor with specified attributes.

The RCSA model described in chapter 7 was used to analyse selected road corridors in Dar es Salaam. The selected road corridors included Morogoro, Nyerere, Bagamoyo, Kilwa and Mandela roads. As mentioned in chapter 3, the reason for choosing these roads was because they were expected to represent the study area.

In using the RCSA model, video footage was first taken at the road mentioned above. The footage was then studied and different scales of classification of urban infrastructure features were entered as inputs for the RCSA model. Such inputs included road corridor features like roadway infrastructure elements, underground infrastructure elements and roadside infrastructure elements.

Via the user interface, the RCSA model was launched, inputs were fed in, and the RCSF indeces for various points on the roads were generated automatically. Indeces were then plotted as shown in Figure 8.1.

Figure 8.1
Results from road features in the selected routes

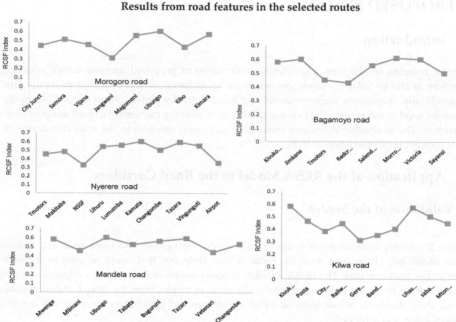

Source: Field data survey (2007)

8.2.2   Data for RCSA Model Validation

Validation of the RCSA model was done by comparing accident data of two selected road corridors, that are Morogoro and Bagamoyo roads with accident data from the Tanzania Traffic Police Department (see Annex II). The road corridors were analysed and used to validate the RCSA model. Analysis of road traffic accident data shows that they were in harmony with model results. The locations along the road corridors found to have the highest RCSF Index also had the highest accident rates.

8.2.3   Results and Implications

Figure 8.1 shows result of data analysis. In Figure 8.1, the RCSF index varies depending on features at different locations, and on the values of ratings assigned intuitively by users (engineers, planners and decision-makers). The higher the value of RCSF index the more risky the location is. The engineers, planners and decision-makers can then look back at features used as the basis for computing the RCSF index and take appropriate counter measures. In Figure 8.1, RCSF indexes are relatively higher along Morogoro road and Bagamoyo road, which means that movement and safety in the road corridor is relatively lower along these roads. It can also be noted that the most critical locations are Ubungo, Mwenge, Changómbe and Gymkana. Countermeasure priorities must be given to these locations.

Figure 8.2 shows results from a different analysis in which RCSF indexes are plotted against distance from reference points along Bagamoyo and Morogoro roads. As the RCSF index graph goes up, the higher the risk of the location, the lower safety level it becomes, and vice versa. This helps the decision-makers to implement countermeasures to enhance safety on those particular road locations.

Figure 8.2
Application of the RCSA model

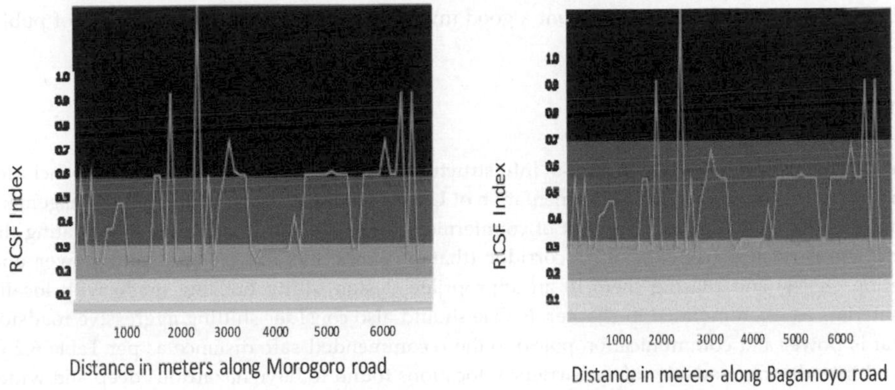

Distance in meters along Morogoro road

Distance in meters along Bagamoyo road

Source: Field data survey (2007)

8.3   Movement and Safety Proposed Plans

8.3.1   Countermeasure Costs

Investment (Plans) costs are categorised by area type either urban or semi-urban cost. Countermeasure costs for this study were estimated based on existing data collated.

(i)   Economic Cost of Deaths and Serious Injuries
Determining the value of a human being has several components, but the basic cost is derived from the expected output of each individual in society. It includes loss of resources and workforce. A related factor is loss of welfare and close relatives may be trapped in poverty if the source of living becomes impossible. Even in less serious cases, depression of losing the beloved may make working impossible for some time. Material losses include a destroyed vehicle, road devices, sometimes the road pavement, but also the driving costs of emergency vehicles and medicine. The methodology used to estimate the economic cost of a road death and a serious injury is based on research undertaken by McMahon and Dahdah (2008). The key equations used are:

- The economic cost of a death is estimated to be: 100 x Gross Domestic Product (GDP) per capita (current price)

- The economic cost of a serious injury is estimated to be: 0.25 x economic cost of a death.

(ii)   Movement and Safer Road Investment Plans

Affordable and economically sound countermeasure plans were considered for road corridor safety improvement. These road corridor improvement options ranged from roadway, underground infrastructure to roadside infrastructure improvement. The plans were produced in the following three key steps:

- Drawing the proposed improvement see the topologies and cross section (attached in Annex III).
- For each section of the road corridor, countermeasure options were tested for their potential to reduce deaths and injuries. For example, a section of a road that has a high level of risk for pedestrians might be a candidate for a pedestrian refuge, pedestrian crossing or signalised pedestrian crossing; and
- Each countermeasure option was assessed for economic effectiveness using a threshold benefit-cost ratio. This ensures that the plans represent a good investment return and responsible use of public funds.

8.3.2   Countermeasures

It is considered necessary to integrate the urban infrastructure development and management. It includes coordinating, planning, designing and implementation of UI projects, and also coordinating management and operations in the road corridors. As part of countermeasures, one should considered relocating the underground infrastructure from the road corridor (that is water and sewer pipes and power and communication cables) and placing them in an appropriate design utility housing made with locally available materials as recommended in chapter 6. One should also consider shifting aggressive roadside features (that is power and communication poles) to the recommended safe distance as per Table 6.2 in chapter 6, and introducing roadside safety barriers at locations found to have hazardous deep and wider open ditches.

Identified countermeasures within the road corridor are shown in Table 8.1. It can be established that by investing a total of TZS 567 million on the narrated countermeasures, 76,400 deaths and serious injuries could be prevented and value of benefit totalling TZS 1,817 million in respect of movement and safety could be achieved

Table 8.1
Proposed countermeasures

Countermeasure type	Length (Km)	Estimated cost (20 years) TZS (Million)	KSI saved (20 Years) TZS	Benefits (20 Years) TZS (Million)	Cost per KSI saved (TZS)	BCR
Relocation of utilities	55	243	17,600	419	1378	17
Redesigning hazardous locations	25	35	10,300	246	748	70
Road surface upgrading	55	72	10,000	237	430	33
Roadside safety barriers	15	20	1,400	34	1384	17
Shoulder widening	21	143	26,100	619	524	43
Delineation	35	6	500	12	397	20
Pedestrian footpaths	28	19	4,200	100	465	53
Regulating roadside commercial activities	22	1	300	7	233	70
Monitoring of UI operations	94	28	6,000	143	274	51
TOTAL	350	567	76,400	1,817	5,833	32

KSI – Killed and Serious Injuries; BCR – Benefit Cost Ratio
Source: Field data survey

The total estimated cost of the movement and safety investment plan is TZS 567 m, KSI saved is 76,400, Total value of safety benefit is TZS 1, 817 million and Benefit Cost Ratio is 32. The investment returns in 7 to 9 months in average, but many individual investments will pay back in 0.5 to 2 months. Table 8.2 shows the investment estimated and saving amount after the implementation of the countermeasures on each road corridor studied. Also the table shows the number of accidents estimated to be reduced if countermeasures earmarked are implemented.

Table 8.2
Annual cost savings

Study Site	Investment	Total 3 years accidents	Accident reduction Number per Year		Annual Saving in Cost Per Year (TZS.)	
			Min. Estim.	Max. Estim.	Min. Estim.	Max. Estim.
Morogoro Road	3162	502	241	291	3321	3969
Bagamoyo Road	2754	387	123	156	2901	3295
Nyerere Road	1178	185	79	107	760	1028
Mandela Road	1820	256	119	154	1692	2364
Kilwa Road	569	233	136	162	1712	2005

Source: Field data survey

8.4   Cost Implications of the Integrated High Performance Infrastructure

This part presents the extent of cost-effectiveness attained by application of IHPI within the urban infrastructure. It presents cost-benefit analysis of the application of IHPI alternatives by using data obtained from previously carried out projects and operational costs associated mainly, with lack of integration in infrastructure development and management of urban utility networks. The potential benefit for improving movement and safety is shown in the Table 8.3 and other cost-saving items are listed hereunder:

(i)   Reduction in property restoration costs due to elimination of open cut for access to physical facilities for UI operation and maintenance;

(ii)   Elimination of utility relocation costs;

(iii)   Elimination of service interruption costs associated with destruction of each other's network due to lack of integrated utilities network database; and

(iv)   Increase of urban space utilisation ratio due to reduction of prolonged encroachment of road corridor surface.

Table 8.3
Improvement requirement and impacts of countermeasures

Black spot	Recorded Accidents	Estimated Accidents	Accident Cost/yr (TZS.)	Improved Cost (TZS.)	Accident Reduction (%)	First Year B/C Ratio (%)	Payback period (Months).
Samora	12	21	300	29,000	47	30	4.5m
Magomeni	23	23	320	18,000	32	70	1.7m
Ubungo	18	31	546	33,000	44-56	110	1m
Kimara	27	30	301	32,000	44-56	50	2.5m
Uhuru	33	38	211	102,000	25-40	16	7-8m
Kamata	49	30	573	136,000	37-45	14	9m
Changombe	28	24	387	186,000	87	18	7m
Tazara	26	21	372	36,000	22-30	31	4m
Vingunguti	25	16	337	32,000	40-45	41	3-4m
Mwenge	30	42	1030	1,850,000	48-59	15	80m
Tabata	24	34	254	10,000	35-51	70	1-4m
Buguruni	13	24	250	14,000	55-63	75	1m
Gymkana	43	31	317	190,000	70	9	15m
Moroco	20	25	225	17,000	24-32	29	3-6m
Uhasibu	6	11	136	12,000	52-64	80	1-2m
Sabasaba	10	15	253	146,000	74-79	17	6-9m
Manzese	22	23	180	35,000	49-57	30	4m
Victoria	12	25	262	119,000	26-41	13	9m
Sayansi	28	18	601	23,000	58-67	89	<1m
Kibo	14	15	212	67,000	59-63	20	5-7m
Extenal	31	36	359	75,000	63-67	32	3-4m
TOTAL/average	543	563	7999	3,569,000	50	46	9

Source: Field data survey (2007)

8.5 Potential Safety Benefits of Integrated High Performance Infrastructure

Evaluation scenarios of the integrated high performance infrastructure elements considered effectiveness of elements of integrated urban infrastructure development, operations and management into improving condition of road features which in turn such features positively improve movement and safety .

8.5.1 Integrated High Performance Infrastructure Concept

The RCSA model was used to show influence of IHPI on movement and safety. Explanatory variables of IHPI influence movement and safety indirectly through their effects on road features, which are road surfaces, road geometry, road drains, traffic control factors and traffic factors.

Generally, the model is posited to show that provision of IHPI to the existing roadway environment will lead into attaining a high level of movement and safety through their positive impact on the road corridor features. Table 8.4 shows elements of Integrated High Performance Infrastructure.

Table 8.4
Elements of integrated high performance infrastructure

No	Variable name	Variable element	Variable definition
1	UI Network layout	Existence of explicit utility zone Consistence of UI network layout	Provision of explicit passage layout of UI Network within the road corridor
2	Integrated UI passages	Presence of Integrated UI passages	Provision of common structures for accommodating utility facilities along and across the road network
3	Synchronisation of standards	Design standards, material standards, layout standards	The synchronised standards of UI networks such that development of UI Networks are implemented together in the project
4	Coordination of UI network service providers	Realising synergy	To link UI network providers for better management and operation in the road corridor
5	UI network database	Accessible data base Adequate data base	Presence of reliable UI network database and sharing information of road corridor
6	Best engineering practices	Compliance to strength Compliance to safety	Engineering technique used to minimise impact to service delivery when carrying out operation activities
7	Monitoring and Evaluation	Minimised potential of conflict	This is a day-to-day activity carried out to insure effective management of road corridor

Source: Field data survey

8.5.2   Integrated Infrastructure Development

Development of urban infrastructure is integrated to maximise the performance level of the urban infrastructure network and reduce adverse effects resulting from the current trend of developing and managing urban infrastructure networks. Features of urban infrastructure networks to be integrated are planning, designing and construction during implementation of development projects.

Development of an integrated urban infrastructure network takes into consideration execution of three measurable elements namely integrating urban infrastructure network layout, integrating urban infrastructure passage facilities and synchronisation of standards as shown in Figure 8.3.

Figure 8.3
Conceptual illustration of integrated UI development

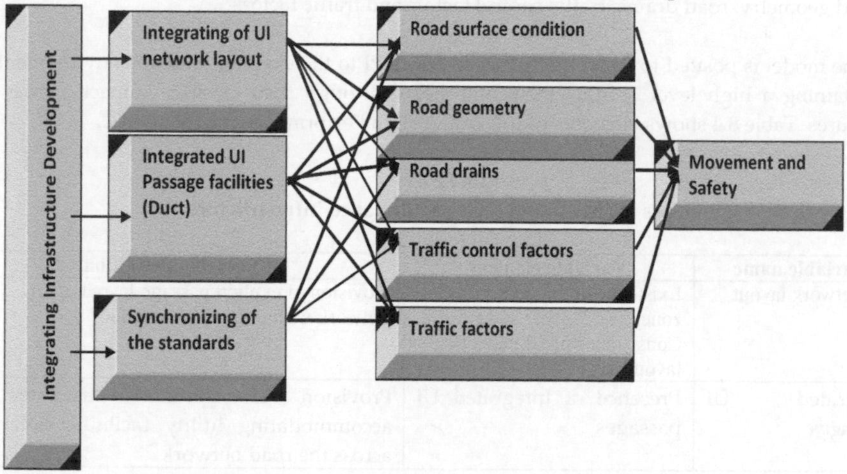

Source: Field data survey

In testing the extent to which such elements improve movement and safety, the benefits that are generally obtained from these elements were related to their likely effects on road features. In turn, resulting state of road features was simulated to yield corresponding levels of movement and safety.

The considered potential benefits of integrating urban infrastructure network development are zoning of road corridor user routes, provision of explicit passages layout of urban infrastructure network in the corridor and synchronisation of urban infrastructure standards. They lead to minimised self-conflicts of different urban infrastructure network, reduce risk of damage of urban infrastructure network, prevent services interruption, minimise interferences to traffic flow, minimise potential risk and reduce space due to encroachment.

8.5.3   Integrated Operations and Management

The integrated high performance infrastructure can be fully attained by integrating development operations and management of urban infrastructure network. Elements of integrated operations and management are coordination corridor utilisation, creation as well as sharing information, undertaking the best engineering practices and setting up of monitoring including evaluation of the activities as summarised in Figure 8.4, which are undertaken in the road corridors.

Figure 8.4
Conceptual illustration of integrated operations and management

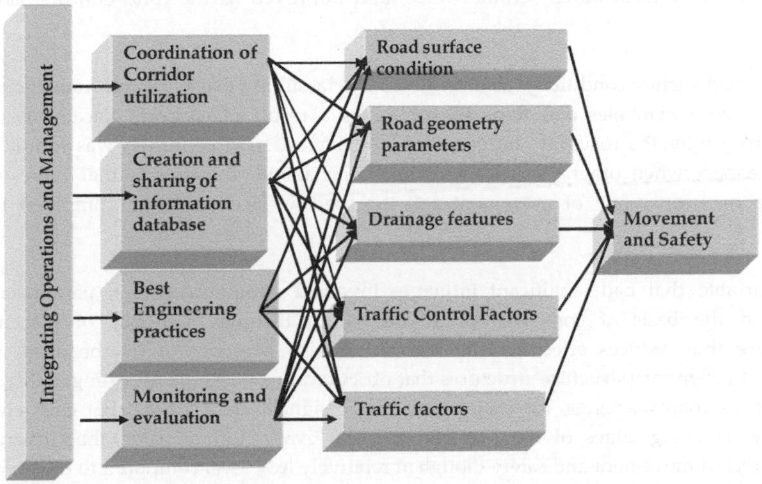

Source: Field data survey

An integrated operation and management system enhances urban infrastructure network monitoring, evaluation and regulates service providers in the road corridor. Their potential benefits are minimised frequencies of interruption to traffic flow, reduced delays, decreased number of conflicts, and improved urban infrastructure network database and information sharing. It also enhances coordination among urban infrastructure service providers, which minimises risks that lead to improving movement and safety.

8.6 Analysis of Evaluation Results

Evaluation given in this section comprises two parts. The first part presents results from an evaluation of model variables of UI that does not include provision of IHPI. In other words, these models simulate the existing condition before provision of the IHPI. The second part gives the results from the model evaluation with provision of the IHPI variable, which shows that the IHPI elements have significant influence on the level of movement and safety when applied to the urban infrastructure in the road corridor.

8.6.1 Evaluation of Existing UI Elements with Road Features

Evaluation of existing urban infrastructure elements with road features used existing urban infrastructure elements and elements of road surface condition, road geometry, road drainage, traffic factors, and traffic control factors to establish their influence on the level of movement and safety thus, present the existing situation of the road corridor without provision of IHPI.

Evaluation of the existing road corridor safety is shown in Figure 8.1. The RCSA model has shown how the urban infrastructure elements affect safety. Results from evaluation of the model showed overall safety without application of IHPI and comprised high RCSF index values, which indicated low level of movement and safety. It was also revealed that movement and safety were impaired as the condition of parent variable of road features became worse, and improved as the condition of road features were improved.

The state of road surface condition was determined by classifying existence and frequencies of occurrence of potholes, open manholes and weak-subsurface as a result of random placement of underground infrastructure under the roadway. Influence of the state on road condition was found to be the most influential aspect when other variables remained unchanged. It suggests that there was substantial relationship between safety of movement and the road surface condition compared to other parent variables.

Another variable that had significant influence involved geometric design parameters. These were classified on the basis of constrictions of lane width through tendency of placement of urban infrastructure that reduces effective lane width and also presence of overhead as well as surface protrusion of urban infrastructure structures that obscured line of sight and riding ability. Traffic control factors such as road markings, road signs and traffic signals are features that exist either within the carriageway or along edges of the carriageway. The evaluation revealed that urban infrastructure elements affected movement and safety though at relatively low level compared to other parent variables.

General evaluation results showed road surface condition, geometric design parameters, traffic factors and traffic control factors to have influence on movement and safety in the respective order of significance. Among the factors that were found to have more influence were those related to existence of mal-placement of urban infrastructure elements within the road corridors.

8.6.2   Evaluation of IHPI Elements with Road Features

(i)   Evaluation of IHPI Elements with Road Surface Condition
Road surface is the structure above the formation design to spread loading over the base and sub-base. Evaluation of IHPI elements with the road surface condition revealed significant relationship to attain high level of safety of movement. Evaluation of results revealed that IHPI was essential for road surface condition improvement, which in turn enhanced movement and safety.

Elements which showed significant relationship to surface condition were underground infrastructure placement, road side infrastructure, pavement design, leakages of pipes, the best engineering practices, operation and management. The road surface condition was found to be affected by these factors and the desired road surface condition was attained by varying these variables.

As it is shown in Figure 8.5, safety levels also vary with the road surface condition. Where the road surface condition becomes poor the level of safety of movement decreases and vice versa. It also indicates the relationship of the level of safety of movement with urban infrastructure condition. The level of safety of movement increases with increasing good condition of urban infrastructure network and the safety of movement is low when these factors are in poor condition.

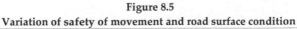

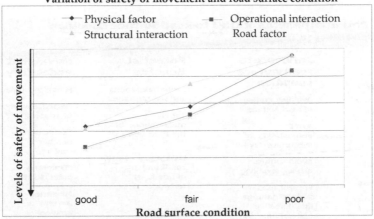

Figure 8.5
Variation of safety of movement and road surface condition

Source: Field data survey (2007)

Furthermore, the condition of the underground infrastructure within the road corridor contributes to deterioration of road surface condition, which in turn affects movement and safety. Results also showed that deterioration of surface conditions was higher at sections where networks of urban underground infrastructures were more congested and this contributed to decline level of movement and safety in the road corridor. Figure 8.6 exhibit the potential benefits of IHPI to road surface condition.

Factors of underground infrastructures which affect movement and safety were reversed by application of IHPI through integrated urban infrastructure passage facilities, which prevent the need for cutting the road surface for underground utility repair. Provision of explicit passages of urban infrastructure network along the road corridor was proposed to be designed to suit the environment with available local materials. Similarly, provision of integrated urban infrastructure passage facilities have the potential of availing easy services to the utilities, preventing overcrowding of corridor of utility infrastructure, prevent cutting of the road surface, prevent vandalism, and minimise operation as well as development cost and avoid utility relocation services during road construction.

An improved level of coordination and communication among stakeholders enhances coordination of activities undertaken by urban infrastructure authorities within the road corridor which minimises damage to road surface and improves urban infrastructure level of service. The IHPI application in the road corridor by providing urban infrastructure network layout, integrated urban infrastructure passages, synchronisation of standards, coordination of urban infrastructure network service providers, urban infrastructure network database, the best Engineering practices and Monitoring and Evaluation, lead to sustainable improvement of road surface condition. Hence, minimised damage to road surfaces, increased effective space of the surface and reducing slippery on the road surface. Eventually, overall road corridor movement and safety in the road corridor are improved.

Figure 8.6
Potential benefits of IHPI to road surface condition

Potential Benefits of IHPI to Road Surface Condition			
IHPI Elements	**Influence to road surface condition**	**Potential Benefits**	**Movement and Safety**
UI Network layer out	Avoiding cutting of road surface	Minimizing roads surface damages	Minimized delay
Integrated UI passages	Avoid deterioration of subsurface layers	Timely maintenance of the road surface	Decrease number of accident
Synchronization of standards	Avoid surface quality impairing	Minimizing of self conflicting of different UI Network	Elimination number of conflict
Coordination of UI network service providers	Prevent road surface damage due defective underground UI	Attaining economy of space	Minimizing potential risk
UI network database	Properly design road surface	Attaining synergy	Minimized severity of accident
Best Engineering practices	Synchronization of UI activities	Easy access to data	Minimized delay
Monitoring and Evaluation			Improved Movement

Source: Survey results 2008

(ii)   Evaluation of IHPI Elements with Geometric Design Parameters
Geometric design of a road is specification of geometry elements of a road. In a way of realising, effectiveness of IHPI into enhancing sustainable safety of movement through their impact on road geometry, provision of IHPI was simulated with geometric design parameters. Geometric design parameters evaluated for their effect to movement and safety included number of lanes, lane width, shoulder width, presence or absence of median, sight distance and cross sectional elements. Results of the evaluation are given in Figure 8.7.

Figure 8.7
Variation of Safety of Movement with the Lane Width

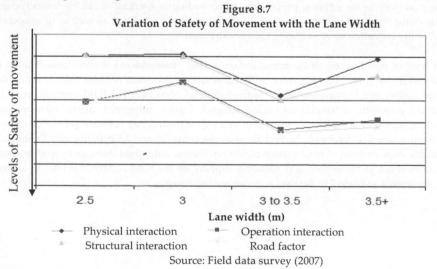

Source: Field data survey (2007)

Overall condition of road geometry appeared to improve as the state of all simulated road geometric parameters improved, with provision of IHPI. Variation of the geometric design parameters and safety is shown in Figure 8.7 where each variable was varied to establish their effects on safety in the road corridors. Analysis of contribution of lane width to safety of movement is exhibited in Figure 8.7. It is shows that there is an optimum range of lane width wider or narrow that result into impairing movement and safety.

In addition, shoulder width exhibits the same trend in Figure 8.8 that is the wider the shoulders the lower the safety of movement. Furthermore, when the shoulder surfaces are not paved and there is shoulder drop-off road safety gets lower.

Figure 8.8
Variation of safety of movement with the shoulder width

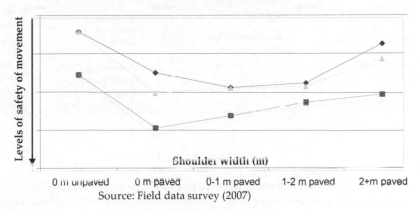

Source: Field data survey (2007)

--◆-- Physical interaction   --■-- Operation interaction
  ▲ Structural interaction   Road factor

Results also revealed that safety of movement on divided roads was lower than that on undivided road sites. These results show potential benefits of provision of the median. On divided roads, pedestrians have the opportunity to cross the road in two stages getting a refuge-island in between opposite traffic streams. Consequently, it increases safety of movement. The number of conflicts is reduced because opposing traffic streams are separated by the median.

The IHPI application in the road corridor through provision of urban infrastructure network layout, integrated urban infrastructure passages, synchronisation of standards, coordination of urban infrastructure network service providers, urban infrastructure network database, the best engineering practices and monitoring and evaluation have potential improvement of geometric design parameters through maintaining effective lane width by reduction of encroachments, maintaining cross falls and sight distance. Figure 8.9 summarises the potential benefits of IHPI to design parameters.

Figure 8.9
Potential benefits of IHPI to geometric design parameters

Potential Benefits of IHPI to Geometric Design Parameters

IHPI Elements	Influence to the geometric design parameters	Potential Benefits	Potential benefits of movement and safety
UI network layout			
Integrated UI passages		Minimising unsafe manoeuvring	Minimised delay
Synchronisation of standards	Maintains effective lane and shoulder width		Eliminated number of conflicts
Coordination of UI network service providers	Attaining stability of movement		Minimised potential risks
UI network database	Maintaining sight distance	Reducing flooding to the road section	Minimized severity of accident
Best Engineering practices	Retaining of cross falls		Improved movement
Monitoring and		Minimising obstruction to line of sight	

Source: Model results

(iii)   Evaluation of IHPI Elements with Traffic and Traffic Control Factors

Traffic factors that were considered in the evaluation model were traffic volume, traffic flow and traffic speed. Other traffic control factors which were considered in the evaluation model were road signs, road marking and traffic signals, which were designed to control traffic in the road corridors. The most important explanatory variable of road corridor safety, which measures the amount of exposure on a given road, is vehicle travel. Results revealed that movement and safety were affected where traffic factors and traffic control factors were impaired.

Higher interaction of pedestrians and motorised traffic is one of the main characteristics of road corridor safety; safety of movement is directly related to the degree of pedestrian exposure to the motorised traffic. Categories of high pedestrian traffic volume due to frontage and neighbouring land uses are used to examine their relationship. In the study area conflicts of pedestrians to motorised traffic were found to be significantly higher than in non-built up areas.

Evaluation results showed that areas prone to emerging unregulated activities that generated higher traffic and also that enchroached the road corridor were associated with unsafe movement. The conflicts increased at an area where land use attracted mixed up traffic movements with different mobility characteristics without corresponding to road provisions. The IHPI, which enhances UI development and operation and management, will lead to improvement of traffic factors and traffic control factors, which lead to increased level of road corridor safety. Figure 8.10 exhibits potential benefits of IHPI to traffic control factors.

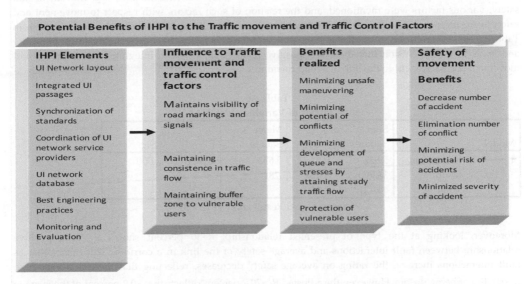

Potential Benefits of IHPI to the Traffic movement and Traffic Control Factors

IHPI Elements	Influence to Traffic movement and traffic control factors	Benefits realized	Safety of movement
UI Network layout		Minimizing unsafe maneuvering	**Benefits**
Integrated UI passages	Maintains visibility of road markings and signals	Minimizing potential of conflicts	Decrease number of accident
Synchronization of standards			Elimination number of conflict
Coordination of UI network service providers	Maintaining consistence in traffic flow	Minimizing development of queue and stresses by attaining steady traffic flow	Minimizing potential risk of accidents
UI network database			Minimized severity of accident
Best Engineering practices	Maintaining buffer zone to vulnerable users	Protection of vulnerable users	
Monitoring and Evaluation			

Source: Based on the model results

Evaluation of the integrated high performance infrastructure elements in the model revealed the strength of the relationships between model variables and safety of movement. Evaluation results also showed the safety benefits realised from integration of urban infrastructure network development, operation and management.

8.6.3 Hypothesis Testing

Simulation results from the proposed model contained negative interactions of the infrastructure existing within the road corridors. However, to ascertain the significant of the results, research hypothesis three was tested.

Hypothesis three states that "the combination of integrated development and urban management as well as integrated road corridor infrastructure is a necessary condition for ensuring enhanced movement and safety in the road corridor in Dar es Salaam."

Variables that constitute the integrated approach, that is, integrated development and urban management as well as integrated road corridor infrastructures were tested if they had any significant relationship with variables of movement and safety. The following are null and alternative hypothesis used for testing.

Null Hypothesis: There is no positive relationship between integrated approach and enhanced movement and safety in the road corridor.

Alternative Hypothesis: There is a positive relationship between integrated approach and enhanced movement and safety in the road corridor.

When interviewing stakeholders on what had led to increased fault interactions and movement and safety, various factors were mentioned, and the relation of such factors with respect to movement and safety were regressed. Looking at the model test of regression, the data deserved the test as shown in Table 8.6 as it is significant. It implies that the null hypothesis is rejected at 5 percent level of significance and leads to the conclusion that there is a positive relationship between integrated development and movement and safety within the link in a road corridor.

Table 8.6
Analysis of Variable (ANOVA)

Model		Sum of Squares	Df	Mean Square	F	Sig.
1	Regression	5.632	1	5.632	20.036	.000
	Residual	38.790	138	.281		
	Total	44.421	139			

Source: Model Result analysis (2008)

Moreover, looking at the type of presented relationship, R=86 percent shows a strong positive relationship between fault interactions and average safety of the link in a corridor. This means that as fault interactions increase the rating on average safety decreases, reflecting that there is a very high possibility of being unsafe. However, the adjusted R^2=70.2 percent reflects that 70.2 percent of the average safety of movement is caused by integrated management of the urban road corridor. The remaining 29.8 percent suggests also that there are other significant factors that contribute to unsafe movement in urban infrastructure including integrated road corridor infrastructures. Furthermore, observing at one coefficient (road aspect) and the constant of the relationship were observed to be significant as illustrated in Table 8.7.

Table 8.7
Analysis of coefficients (a)

Model		Un-standardized Coefficients		Standardized Coefficients	t	Sig.
		B	Std. Error	Beta	B	Std. Error
1	(Constant)	3.059	.406		7.537	.000
	Forced relocation; uncoordinated management; lack of monitoring and evaluation; lack of maintenance; lack of database	.006	.085	.007	.068	.000
	Lack of report procedure	.096	.090	.107	1.063	.002
	Lack of master plan	.096	.089	.101	1.078	.0083
	Lack of coordinated standards	.126	.093	.138	1.347	.00180
	Poor urban management and land use	.000	.092	.001	.005	.00996

Source: Model results analysis (2008)

8.7   Conclusion

Evaluation results revealed the existence of a relationship between movement, safety and independent variables used in the model. Influence of road surface condition, road geometry, traffic factors and traffic control factors on movement and safety within the road corridors were realised at different conditions of variables under consideration. Furthermore, they were noted to be more beneficial to movement and safety when they were an incumbent part of the integrated approach in road corridor.

The simulated variables of IHPI elements in the model showed a significant relationship to levels of movement and safety. Variations of an evaluation results for the urban infrastructure network with and without IHPI. The percentage of reduction of potential risks was calculated by taking RCSF index of the existing road corridor features minus the RCSF index of proposed road corridor over RCSF index of the existing road corridor features times 100 percent. Results are presented in Table 8.8.

Table 8.8
Summary of evaluation results

Proportion of good condition of road features		
S/N	Road features	RCSF index Score before provision of IHPI	RCSF index Score after provision of IHPI
1	Road surface condition	41.9 percent	94.8 percent
2	Geometric design parameters	45.5 percent	86.1 percent
3	Drainage features	38.6 percent	82.7 percent
4	Traffic Factors	39.4 percent	93.1 percent
5	Traffic control factors	48.2 percent	85.5 percent
6	Safety of movement	42.2 percent	90.6 percent

Source: Field data survey (2007)

As shown in Table 8.8 there is substantial influence of IHPI elements on the condition of road features. Improving quality of individual road feature elements appears to improve the state from poor to good. Improvement of road feature conditions lead to enhanced road corridor movement and safety. As the road features become good, the level of movement and safety is also improved. The RCSA model served as a way of expressing the potential benefits from provision of an integrated high performance infrastructure to safety of movement in the road corridor. The different levels of safety of movement were attained by varying model variables so as to evaluate their influence on safety of movement.

Results from the analysis infer not to reject hypothesis three that the combination of integrated development and management as well as integrated road corridor infrastructure is a necessary but not sufficient condition for ensuring enhanced movement and safety in the road corridor in Dar es Salaam.

PART FOUR: SYNTHESIS

This fourth and final part of the thesis discusses research findings, draws a number of conclusions and gives recommendations for way forward on researched issues. It contains chapter 9 which presents a synthesis comprised of discussion of results, conclusions and recommendations.

9

9.0   SYNTHESIS

This chapter discusses the extent to which outcomes of the conducted research provided answers to research questions, results concerning the formulated hypotheses and the way results relate to the theoretical framework. Finally, some practical conclusions are presented.

9.1   Discussion of the Research Findings

9.1.1   Infrastructure Interdependences in the Urban Road Corridor

The first research question required identifying characteristics of interconnections among infrastructure assets (roadway, utilities, and other facilities) within the urban road corridor and between the assets and the corridor conditions in Dar es Salaam.

All five studied road corridors were found to have infrastructure assets that exhibit geographical interdependence by virtue of being in proximity due to sharing of location within the road corridor. Proximity of infrastructures was identified to significantly contributing as observed physical interactions.

Furthermore, the infrastructure assets were found to link with each other through inflicting physical faults that frequently result into physical deterioration of components of the infrastructure assets. There were significant numbers of faults on infrastructure caused by distresses resulting from malfunctioning of adjacent infrastructures. There existed also logical interactions in form of independent and dependent components whereby the assets operated in a logical connection, such as one is the controller of intelligence and the other is the driven consumer.

The study further found inevitable existence of interdependence among urban infrastructure networks. Such findings support what was revealed by Ferreira and Flintsch (2004) and the need for adopting an integrated approach. The findings significantly support hypothesis one, concerning infrastructure assets (the roadway, utilities and other facilities) in an urban corridor that are geographically interdependent primarily by virtue of their physical proximity as well as operational and faulty interactions among the systems.

9.1.2   Movements and Safety Conditions in the Road Corridor

Identification of effects of existing urban infrastructure interaction on movement and safety was the focus of research question number two. Analysis of collected data revealed that interactions of infrastructure within the road corridor had a significant contribution to impairing mobility and level of safety.

Failure of underground infrastructure such as water and sewer pipes was noted to damage road surface pavement and pedestrian walkways. Such distresses were in the form of ruts, potholes and depressions. Such failures impaired steady movement as they affected road surface quality. This led to conflicts which were potentially hazardous to traffic movement.

The analysis also revealed two corridor conditions related hazards, namely urban infrastructure development related hazards and urban infrastructure operations and management related hazards. Each of these potential hazards to movement and safety are discussed below.

9.1.3   Urban Infrastructure Development Related Conflicts

The urban infrastructure development related conflicts were found to arise while implementing activities of infrastructure repair, maintenance, rehabilitation, upgrading and new construction within the road corridor. These activities were more profound under condition of non-integrated practice; this was noted to be significant when simulated under a condition of integrated approach. Results are in line with what Brown (2005), deduced that integration is derived from best practices of planning, designing and construction.

Results in chapter 5 show occurrences of conflicts that originated from urban infrastructure development related hazards and which emanated from unsynchronised standards on works related to development and operation for different infrastructures. The majority of interviewees concurred about the fact that during development of urban infrastructure network, several activities are involved which increase conflicts and hinder traffic movement in the road corridor. There are activities that are brought by lack of integration of the urban infrastructure network, such as relocation of the utilities, excavation of road surface for maintenance of underground utilities crossing and repair of damaged underground infrastructure as detailed in chapter 5. They contravene in attaining desired movement and safety within the road corridor.

It was also found out that there were offending features such as deep open drainage ditches and randomly placed huge and hard utility poles close to the edge of the road. Furthermore, there were unsafe utility placement and manholes within the road subsurface that unsafely protruded to the surface. These were caused by uneven settlements of the two adjoining materials such as road pavement and the manholes. All of these features were contributed by uncoordinated infrastructure operations that also covered both design and implementation of the design features.

It was established that many elements of the urban infrastructure were designed and placed randomly. This was due to lack of proper design records of the layout of underground infrastructure networks and lack of coordination among the operators. The collected data indicated that there were no harmonised regulations and procedures used by authorities managing urban infrastructure. The prevailing standards of depth from the road surface, colours of the pipes, cables and the distances of the poles from the road edge were not properly defined and therefore always compromised.

It is generally known that a work zone poses hazard to road users, when the work zone is left unprotected while the work is not in progress (Ogden, 2000). However, inventory data from infrastructure construction and repair activities in the road corridor were seen to affect movement and safety. A number of road users such as pedestrians, cyclists, motorcyclists and vehicles were frequently bumping and falling in exposed pits due to improper protection and placement of warning signs to road corridor users during construction.

9.1.4   Urban Infrastructure Operations and Management Related Conflicts

There were operations and management conflicts faced by infrastructure users that originated from routine operations and management. The study established the fact that urban infrastructures in the study corridor had regular operational activities that individually created vulnerability to movement and safety. While in service, urban infrastructure undergoes wear, tear and aging; which inevitably require both preventive and reactive measures. All such measures on the infrastructure bring about conflict of traffic movement and therefore, movement and safety are jeopardised.

Among the observed urban infrastructure operation and management related hazards, were damages on road surface and walkways. It was established that damages were mainly due to operational and management aspects of the road corridor and its neighbourhood. Results also showed frequent occurrence of uncoordinated utility placement within the road corridor, prolonged infrastructure repair as well as suspended restoration of surface cutting related to repair of underground utility network facilities.

Other noted features were related to socio-economic issues. There was rampant encroachment on the road corridor that affected movement and safety. Poor management of road corridor eventually forced infrastructure agencies to install network facilities in an ad hoc basis. There were no master plan indicating proper usage of the road corridor and location of the infrastructure network. Such situation resulted in a multiplicity of activities in the road corridor that eventually affected movement. Furthermore, the prevailing lack of effective management of the road corridor also led to incidences of frequent encroachment of road corridor space caused by activities emanating from adjacent land usage and infrastructure repair works that had contributed to mal-functioning of the existing urban infrastructure network.

It was also noted that cross cutting issues that went beyond the urban infrastructure network affected operation and management of infrastructure. Changes in land use patterns created excessive demands for infrastructure services beyond the existing capacity. Given such a situation, emerging users scrambled for limited services making uncoordinated connectivity that significantly led to destroying infrastructure networks (like land use close to a business centre or industrial area which does not match with the infrastructure design capacity).

The findings further more show significant contribution of the road surface conditions to the level of movement and safety. It was also revealed that as the road surface condition worsens so does the level of movements and safety. Thus overall, the study results did not have significant statistical inference to reject the hypothesis that typical urban road corridor in Dar es Salaam affects movement and safety primarily due to operational and faulty interactions among infrastructure assets in the road corridor as well as the approach to development and management of the assets.

9.1.5 Integrated Infrastructure for Sustainable Improvement of Movements and Safety

Findings in chapters 4 and 5 call for determination of approaches capable of mitigating detrimental interactions in order to attain sustainable improvement of movement and safety within the road corridor. Undesirability brought by non-integrated urban infrastructure makes it apparent that in order to attain sustainable improved level of movement and safety, an IHPI is to be adopted to cater for cross-cutting issues of coordination in urban infrastructure development, operations and management.

The IHPI is an infrastructure system that is built-up by proven best planning, design and management practices. This includes scheduling and coordination of disparate activities in order to minimise impacts on citizens and businesses as pointed out by Vogel (2006).

The Road Corridor Safety Analysis (RCSA) model was applied to assess the potential benefits from IHPI and suitability of an integrated approach into improving movement and safety. Simulation of an integrated development approach through provision of high performance integrated infrastructure resulted into attaining significant potential benefits. Integrated layout of infrastructure, usage of integrated structures and synchronisation of standards resulted in an improved movement and safety within the roadway. The attained minimised conflicts from the simulated parameters brought about a more sustainable road corridor and improved handling of traffic flow and safety of movement within the road corridor.

Results also showed significant contribution of other residual parameters that did not emanate from the infrastructure assets. These parameters might be accounted for by cross-cutting issues such as local governance and socio-economic issues, which required integration beyond road corridor related environment. Issues of adjacent land use, human behaviours and socio-economic activities were noted to have the potential for reducing effectiveness of adopting high performance integrated infrastructure if not considered in line with this adoption.

However, the study results did not infer to reject the hypothesis that a combination of integrated infrastructure development and management as well as integrated road corridors is a necessary strategy for sustainable improvement of movement and safety in Dar es Salaam.

9.2   Conclusions

This part presents conclusions of study; the findings revealed that there exists a mutual interdependence of urban infrastructures within the road corridor. The study also established that physical characteristics of urban infrastructure networks and the way they are operated have significant effects in bringing about encroachment of space, inflicting deterioration to each other, obscuring sight and impairing service availability to other users.

There are serious effects induced by interaction of existing infrastructure assets to the road corridor, which in turn pose adverse effects to movement and safety. Such effects are caused by faulty and operation interactions of urban infrastructure in a manner that significantly inflicts deterioration of road surface and impair movement and safety. Such impairing is seriously escalated by mismatches in standards and inferior planning, designs, development and operational approaches of the urban infrastructure.

The RCSA model showed that urban infrastructure interaction affect safety of movement by affecting road elements in the road corridor. The model showed the overall safety of movement without application of IHPI; with indication of safety of movements being increased number of accidents, increased number of conflicts and increased potential hazards. It was found out that safety of movement was impaired, as condition of the parent variable of road corridor features become worse and improving as the road corridor features condition improves.

Influence of the road surface conditions was found to be the most influential when other variables remained unchanged. It suggests that there is a strong relationship between movement, safety and road surface condition compared to other parent variables.

Other variables that had significant influence were geometric parameters. Randomly placed urban infrastructure reduced effective lane and shoulder width. Presence of overhead as well as surface protrusion of urban infrastructure structures obscured the line of sight and deteriorated geometric parameters due to inferior reinstated excavated road surface. Findings showed that road surface conditions, geometric design parameters, traffic factors and traffic control factors had influence on movement and safety in the respective order of significance.

The significant effect of urban infrastructure interaction on movement and safety jeopardises the competitiveness of the Dar es Salaam City. Containment of such interaction which impairs movement and safety will make the city comparatively safer, more capable of handling multiple roadway users, and able to offer high effective utilisation of road capacity due to non-construction of lane width. Furthermore, the city will become more competitive by lowering travelling cost through reduction in travel time offered by steady flow of traffic and ensuring that non-motorised traffic has suitable conditions for mobility.

The preceding discussion proves that in order to attain improved movement and safety, appropriate integrated approaches are required to address the escalating problems of urban infrastructure interaction. The study established that provision of IHPI is an appropriate approach due to its potential benefits in addressing urban infrastructure development and operations management related problems. The IHPI influences the level of movement and safety in terms of their best practices in planning, designing, operations and management. The study also showed the relevance of having the best engineering practices and IHPI for sustainable improvement of movement and safety in the urban road corridor. Thus, it can be concluded that establishment of the IHPI system is an effective approach for attaining sustainable improvement of movement and safety in the road corridor.

9.3   Contribution to Knowledge

This study principally has thrown light on effects of interdependencies of urban infrastructure assets under the prevailing road corridor environment condition on movement and safety during urban infrastructure development, operations and management. It also established interdependencies among infrastructure assets in Dar es Salaam and the relationship between the infrastructure assets, movement and safety in the road corridor.

It was found out that the road corridor in Dar es Salaam is unsafe primarily due to operational and fault interaction among infrastructure assets in the road corridor as well as the approach to development, operation and management. During development, operation and management of urban infrastructure network, several activities were involved, which were in conflict with traffic movement in the road corridor. They contributed to creating unsafe conditions in the road corridor. Furthermore, it was established that as the road corridor condition becomes worse, so does safety of movement.

Results from this research are in valuable for improving movement and safety in Dar es Salaam and in other cities as well as municipalities in Tanzania. They could be used for improving the existing urban infrastructure assets and incorporating such assets during planning, designing, construction and management. It was also found out that adoption of the integrated high performance infrastructure in the road corridor is a necessary strategy for ensuring a sustainable improvement of movement and safety of people and vehicles in the road corridor.

One of the main tasks of this research was acquisition of adequate and reliable data with a wide range of potential urban infrastructure, movement and safety variables to attain the defined objectives. The fieldwork was carried out to collect required data as a principal part of the research that provided an essential database for research investigation, which was not available before. Establishing of these databases for the study sites is also a valuable achievement by itself. The data collected in this study could be used to provide ground for improvement of movement and safety in road corridors in Tanzania.

The RCSA model was developed and used for evaluating safety of movement. The case study showed how the urban infrastructure variables affect safety of movement. Results from the case study also showed that overall levels of safety of movement improved after application of IHPI, with indicators such as reduced number of accidents, reduced number of conflicts, minimised severity of accidents and reduced potential risks of safety of movement. It was found out that safety of movement was impaired, as condition of the parent variable of road corridor features become worse and improving as the road corridor features condition improves.

9.4   Recommendations

Several recommendations were drawn from this study, which covered practical implications, recommendations to practitioners, recommendations to policy makers and areas for further studies.

9.4.1   Practical Implications

Practical implications are explained in terms of recommendations to practitioners and recommendations to policy makers. The study found out that there was non-compliance to safety provisions emanating from design in geometric, pavement and drainage features. The study also established the fact that there was deterioration of movement and safety due to uncoordinated development and operational activities by the urban infrastructure operators within the road corridors. This also economically led to misuse of resources used frequently for relocation and repair of damages caused by one infrastructure system to another.

There were also mismatches among developers and operators in adopting standards that defined the bottom line of movement and safety compliance with regard to movements of users within road corridors. Furthermore, there was a gap in defined hierarchy of responsibilities among agencies that controlled and monitored land utilisation within a road corridor and in neighbourhood development.

9.4.2 Recommendations to Practitioners

As a way of reducing the negative interactivity layout of urban infrastructure networks have to be in harmony with each other in terms of design, development and operations for attaining sustainable improved movement and safety. Provision of utility corridors within the road corridor during design is necessary in order to avoid overcrowding of utility infrastructure in the road corridors, prevent cutting of the road network for repair of the underground infrastructure, easy maintenance and minimise relocation cost during infrastructure development.

There is necessity for developing shared urban infrastructure development master plans that will enhance involvement and coordination during implementation of urban infrastructure development, operations and management for sustainable improvement of movement and safety. Economically, that will save resources used for relocation and repairing damage.

It was found out that the road surface condition was directly affecting movement and safety in the road corridors. The urban infrastructure factors were found to affect safety of movement directly. Affecting road surface conditions were underground infrastructure interaction, design, operations, pavement condition and the drainage system conditions. The remedial measure would be through integration of urban infrastructure development, operation and management and use the special utility housing for underground infrastructure such as utility corridor, joint trenching and multi-duct conduit, depending on availability of space in the road corridor as pointed out in chapter 6.

Geometric design parameters considered in enhancing safety of movement in the road corridors included number of lanes, lane width, shoulder width, median, sight distance and cross sectional element. Among these factors, lane width, shoulder width and side slope were found to have significant effects. This calls for upgrading such elements so as to comply with safety of movement through the best engineering practices.

There is a need for utility service providers to seek permission from the road authority, to allocate space for installation of their utilities. On the other hand, the Road Agency should provide estimation costs to the utility service providers for restoration of pavements to their original condition so as to ensure quality.

9.4.3 Recommendation to Policy Makers

In Dar es Salaam, traffic is generated by land use, and location of activity centres influence travel patterns. Therefore, the amount of travel can be minimised and undesirable interaction movements avoided by proper land use planning. The planners and city authority should adopt the best practices in planning and implementing the planned activities on time.

There is a need to have legislation or laws that will facilitate establishment of effective land use planning for adequate road corridors that are able to accommodate the road components, various utilities and keep human activities away from the road corridors.

A digital urban infrastructure which is GIS and GPS based should be established for easy operationalisation and management of underground urban infrastructure. The authorities should develop and establish a comprehensive database for urban infrastructure. This will facilitate effective maintenance and operation of urban infrastructure without affecting the corridor environment and other infrastructure.

Urban infrastructure and road corridor management system should be established for effective development, operation and management of urban infrastructure for sustainable improvement of movement and safety in the road corridor.

The IHPI needs to be adopted so as to cater for cross cutting issues of coordination in planning and implementation, controlling as well as monitoring among urban infrastructure owners and operators.

Guidelines for managing safety of movement in the road corridors should be instituted. This will enhance an integrated and holistic approach to urban infrastructure in planning, programming, implementation, operation and maintenance, and as a result, the level of coordination and communication of stakeholders managing urban infrastructure will be enhanced.

An efficient database of road traffic accidents and recording should be established. The GPS recording system of road traffic accidents should be used by Traffic Police. Adoption of the system will enable road accident reports to contain details of the road sections, and precise locations of the accidents would be reported with coordinates.

 Governmental bodies should dedicate themselves more than before to pave the way towards improving movement and incorporating safety elements at an early stage during planning, designing, construction and maintenance of urban infrastructure network in the road corridor.

9.4.4   Areas for Further Studies

By critically examining achievements of this study with respect to the objectives set earlier, it has been found out that data on road accidents are inadequate. There were no adequate records on secondary data, which recorded accidents in terms of precise location. This makes reporting of fatal and non-fatal accidents uncertain. Therefore, under-reporting of road traffic accidents was observed to be a big problem. The location of an accident is reported broadly by using the name of the area instead of the road name and location name of the scene, with GPS coordinates. Therefore it is very important that the location of an accident be accurately recorded in the accident report in order to identify precise locations of the accidents that relate with safety of movement posed hazards.

Generally, the findings of this study provide directions for further studies in order to create a sustainable improvement of safety of movement in the country, by establishing a sound traffic accident recording system which will be supported by the effective computerised accident data base for precise coding of the traffic incidences along the road corridor.

REFERENCES

Abbas, A. K. (2004): Traffic safety assessment and development of predictive models for accidents on rural roads in Egypt. Accident Analysis and Prevention 36: 149-163.

Abdel Aty, M. A. and Radwan, A. E. (2000): Modelling traffic accident occurrence and involvement. Accident Analysis and Prevention 32: 633-642.

Adams, B. J. and Papa F. (2000): Urban Stormwater Management Planning with Analytical Probabilistic Models. John Wiley & Sons, Inc.

Adams, J. G. (1987): Smeed's Law: Some further Thoughts. Traffic Engineer Control. 28 (2): 70–73.

Adedapo, A. A. (2007): "Pavement Deterioration and PE Pipe Behaviour Resulting from Open-cut and HDD Pipeline Installation Techniques" A thesis presented to the University of Waterloo in fulfillment of the thesis requirement for the degree of Doctor of Philosophy in Civil Engineering Waterloo, Ontario, Canada, 2007

Ahmed, I. and Putcha, S. (2000): Evaluation of Compaction Methods for Pipe Trench Backfill in Areas of High Water Table. Transportation Research Record 1736. Washington DC: Transportation Research Board.

AI (1982): Research and Development of the Asphalt Institute's Thickness Design Manual (MS-1). 82-2, Asphalt Institute.

Akinyemi, E. O. and Medani, T. O. (1999): A Simulation Model for the Effects of Motorcycles on Performance at Sgnalized Intersection Approach Sections. International Institute for Infrastructural, Hydraulic and Environmental Engineering. IHE Working Paper TRE 026.

Alfacon, P. (1989): Integrated Urban Infrastructure Development Program for WEST JAVA, Asian Development Bank, West Java.

Al-Masaeid, H. R. and Suleiman, G. M. (2004): Relationships between urban planning variables and traffic crashes in Damascus. Road and Transport Research 13(4):63-73.

Al-Suhaibani, A. Alani, B. A. and Al-Kharashi, I. A. (1992): Effects of utility cut patching on pavement deterioration. Journal of King Saud University; Engineering Sc. Vol. 4, no2, pp. 171-192.

AMEC (2002): Evaluation of Pavement Cut Impacts, prepared for League of Arizona Cities by AMEC Earth & Environmental, Inc. Phoenix, Arizona.

American Association of State Highway and Transportation Officials (AASHTO) (2005): A Policy on the Accommodation of Utilities within the Freeway Right-of-Way, Washington, DC;

American Association of State Highway and Transportation Officials (AASHTO) (2005): A Guide for the Accommodation of Utilities within the Freeway Right-of-Way, 2005, Washington, DC.

American Public Works Association (1997): Managing Utility Cuts. Kansas City, MO.

American Association of State Highway and Transportation Officials (AASHTO), (2005): A Policy on the Accommodation of Utilities within the Freeway Right

Amini, F. (2003): Potential applications of dynamic and static cone penetrometers in MDOT pavement design and construction. FHWA/MS-DOT-RD-03-162. Jackson, MS:

Department of Civil Engineering, Jackson State University.

Amis, G. (1996): An Application of Generalised Linear Modelling to the Analysis of Traffic Accidents. Traffic Engineering Control. 37(12): 691-696.

Andreassen, D. C. (1985): Linking Deaths with Vehicles and Population. Traffic Engineering Control. 26 (11): 547-549.

Andreassen, D. C. (1991): Population and Registered Vehicle Data vs. Road Deaths. Accident and Prevention 23 (5): 343-351.

APWA (1991) Recovering Cost of Pavement Cuts, Proc. Project Sponsors 1991 Annual Meeting, American Public Works Association, Sept 1991.

APWA (1997) Managing Utility Cuts, Special Report American Public Works Association

Armstrong E. (1978): The quality of urban life. Paper prepared for conference, Manchester, Sept. 1978

Aronson,   D.   (1996)   Overview   of   Systems   Thinking.   [Online] http://www.thinking.net/Systems_Thinking/OverviewSTarticle.pdf [Access date: 18th February, 2008]

Arudi, R. Pickering, B. and Flading, J. (2000): Planning and implementation of management system for utility cuts. Transportation Research Record, 1699. Washington, DC: Transportation Research Board.

Assum, T. (1997): Attitude and Road accident Risk. Accident Analysis and prevention 29, pp. 153-159.

Autret, P. and Brousse, J. L. (2000): VIZIR: Method for the Quality Rating of Paved Roads. Laboratoire Central des Ponts et Chaussées.

Babuska, R. (1996): Fuzzy Modelling and Identification, PhD Thesis, Technical University Delft, The Netherlands.

Bajpai, A. (2006) Millennium Development Goal 7 - Ensure Environmental Sustainability: How are we doing? "Vision", the E-Journal of the WSCSD.

Barabási, A. L. (2002): Linked: The New Science of Networks, MS: Pesseus Publishing, Cambridge.

Berhanu, G. (2004): Models relating traffic safety with road environment and traffic flows on arterial roads in Addis Ababa. Accident Analysis and Prevention 36: 697-704.

Bodocsi, A. Pant, P. D. Aktan, A. E. and Arudi, R. S. (1995): Impact of Utility Cuts on Performance of Street Pavements-Final Report. Cincinnati, OH: The Cincinnati Infrastructure Institute, Department of Civil and Environmental Engineering, University of Cincinnati.

Bong-Min, Y. and Jinhyun, K. (2003): Road Traffic Accidents and Policy Interventions in Korea. Injury Control and Safety Promotion 10, pp. 89-94.

Broughton, J. (1998): Predictive models of Road Accidents Fatalities. Traffic Engineering Control. 29 (5): 296-300.

Burmister, D. M. (1943): The Theory of Stresses and Displacements in Layered Soil Systems and Applications to the Design of Airport Runways. Proceedings, Highway Research Board, 23, pp.126-144.

BWDB (1997): Training Manual on Project Management. Bangladesh Water Development Board (BWDB), Sir William Halcrow & Partners Ltd. BCEOM and DHV consultants in association with development Design Consultants Ltd. Approtech Consultants Ltd.

CEFI (1996): Level Of Investment Study: Facilities and Infrastructure Maintenance and Repair. Civil Engineering Forum for Innovation, Washington, D.C.

CEFI (2007): The Challenges of Change: Applying Innovation and Knowledge to Solve the Nation's Infrastructure Challenges Civil Engineering Forum for Innovation. [Online] http://content.asce.org/cefi/videoclips.html [Access date: 13th September 2007].

Chang, L. and Mannering, F. (1999): Analysis of Injury and Vehicle occupancy in truck and non-truck-involved accidents. Accident Analysis and Prevention 31(5), pp. 579-592.

Chow, C. and Troyan, V. (1999): Quantifying Damages from Utility Cuts in Asphalt Pavement by Using San Francisco's Pavement Management Data. Journal of the Transportation Research Board, No 1655, pp. 1-7.

Clough, P. Duncan, I. Steel, D. Smith, J. and Yeabsley, J. (2004): Sustainable Infrastructure: A Policy Framework. New Zealand Institute of Economic Research (NZIER). [Online] [Access date: 29th March, 2008].

Coates, S. and Sansom, K. (1999): Project Planning and Monitoring. Unpublished Lecture notes, WEDC, Loughborough University, United Kingdom

Convery, F. J. (1998): Challenges for Urban Infrastructure in the European Union: European Foundation for the Improvement of Living and Working Conditions, Luxembourg.

Cowan, H. J. (1953): The strength of plain, reinforced and prestressed concrete under the action of combined stresses with particular reference to combined bending and torsion of rectangular sections. Magazine of Concrete Research, Vol. 5, No. 14, pp. 75-86.

DAG (2004): Drinking Water and Sanitation. Development Alternatives Group (DAG) Newsletter for Human Sustainable Development.

Day, T. J. (2000): Sewer Management Systems. John Wiley & Sons, Inc.

Department of Public Works. (1998): The Impact of Excavation on San Francisco Streets. San Francisco, CA: Department of Public Works, City and Country of San Francisco, and Blue Ribbon Panel on Pavement Damage.

Diamond D. (2002): "JFK Air Train Tour of 7/13/02," NYCRail.com, http://www.nycrail.com/misc/airtrain.htm Accessed 7 April 2004:

Don McKenzie (2005): "Review of Issues Affecting Utilities and Road, Rail and Motorway Corridors" (2005) Submission to Ministry of Economic Development – IPENZ (Institution of Professional Engineers NZ) Transportation Group.

Emery, J. J, and Johnston, T. (1986): Unshrinkable fill for utility cut restorations. Publication SP - American Concrete Institute, 1986, pp. 187-212.

Emery, J. J. and Johnson, T. H. (1987): Influence of Utility Cut on Urban pavement Performance, 2nd North American conference on managing pavements, proceedings Vol. 1, Toronto.

Farshad, M. (2006): Plastic Pipe Systems: Failure Investigations and Diagnosis, Elsevier, 1st.

Farshad, M. and Necola, A. (2004): Strain Corrosion of Glass Fiber-reinforced Plastics Pipes. Polymer Testing, 23(5), pp. 517 – 521.

FDT (2002): Flexible Pavement Design Manual. D. o. Transportation, ed. Florida Department of Transportation, United State.

Federal Highway Administration (2002): "European Right-of-Way and Utilities Best Practices"

U.S Department of Transportation International Technology Exchange Program
   Publication No. FHWA-PL-02-013 August 2002.

Fox, W. F. (1994): Strategic options for Urban Infrastructure Management.
   UNDP/UNCHS/World Bank Urban Management Programme.

Freund, D. M. (1992): Urban & Regional Information Systems Association. URISA Proceedings
   vol. 1.

Fridstrom, L. (1991): An Aggregate Accident Model Based on Pooled, Regional Time-Series Data.
   Accident Analysis and Prevention. 23(5), pp. 363 – 378.

Gadgil, A. J. and Derby, E. A. (2003): Providing Safe Drinking Water to 1.2 Billion Unserved
   People. 96th Annual AWMA, Conference San Diego, CA, Pittsburgh.

Gassman, S.L. Pierce, C.E. and Schroeder, A. J. (2001): Effects of prolonged mixing and
   retempering on properties of controlled low-strength material (CLSM). ACI Material
   Journal 98 (2).

Gaudry, M.(1987): Responsibility for Accidents; relevant Results Selected from the DRAG model.
   Paper Prepared for the Seventeenth Annual Workshop on Commercial and
   Consumer Law, University of Toronto, October 16-17, 1987.

Gharaibeh, N. G. Darter, M. I. and Uzarski, D. R. (1999): Development of a prototype highway
   asset management system. Journal of Infrastructure Systems, 5(2), 61-68.

Ghataora, G.S. Alobaidi, I.M. (2000): Assessment of the performance of trial trenches backfilled
   with cementitious materials. International Journal of Pavement Engineering 1 (4).

Gili, J.A. and Alonso, E.E. (2002): Microstructural deformation mechanisms of unsaturated
   granular soils. International Journal for Numerical and Analytical Methods in
   Geomechanics 26:433-468.

Gleick, J. (1987): Chaos: Making a New Science, Viking, New York.

Golob, F. Thomas and Regan, C. Amelia (2004): Traffic Conditions and Truck Accidents on
   Urban Freeways. Institute of Transportation Studies, Univesity of Calfonia, Irvine,
   Calfonia 92697-3600, U.S.A, pp. 1-35.

Habibian, A. (1994): Effects of temperature changes on water mains break. Journal of
   Transportation Engineering, ASCE, Vol. 120, No 2, pp. 312–321.

Hadi, M. A., Aruldhas, J. Chow, L. F. and Wattleworth, J. A. (1995): Estimating Safety Effects of
   Cross-Section Design for Various Highway Types Using Negative Binomial
   Regression. Transportation Research Record 1500, TRB, National Research Council,
   Washington, D.C. 169-177.

Halfawy, M. M. R., Newton, L. A. and Vanier, D. J. (2006): Review of Commercial Municipal
   Infrastructure Asset Management Systems Information Technology in Construction,
   II (Special issue), 211-224

Hall, R. D. (1986): Accidents at four-arm single carriageway urban traffic signals. Transport and
   Road Research Laboratory, Department of Transport, Southampton University.

Hans Von Holst, A. N. and Andersson, A. E. (1997): Transportation, traffic safety and health.
   Royal Institute of Technology, WHO Collaborating Centre, Sweden.

Hedlund, J. H., Harsha B. and Hutt K. R. (2005): Countermeasures That Work: A Highway Safety
   Countermeasure Guide for State Highway Safety Offices. Information Draft for
   NCHRP 17-33, 5-12-05.

Heller, M. (2001): "Interdependencies in Civil Infrastructure Systems," The Bridge 31 no. 4; Available through the National Academy of Engineering.

Henn, R.W. (2003): AUA Guidelines for Backfilling and Contact Grouting of Tunnels and Shafts. ASCE Press Reston, VA: Technical Committee on Backfilling and Contact Grouting of Tunnels and Shafts of the American Underground Construction Association.

Holtz, R.D. and Kovacs, W.D. (1981): An Introduction to Geotechnical Engineering. Upper Saddle River, NJ: Prentice-Hall Civil Engineering and Engineering Mechanics Series, Prentice Hall.

Howard, A. K. (1977): Modulus of Soil reaction Values for Buried Flexible Pipe. Journal of the Geotechnical Engineering Division, ASCE, Vol. 103, No 1, pp. 33-43.

Huang, Y. H. (1993): Pavement Analysis and Design, Prentice Hall, I, New Jersey.

Humphrey, M. H. and Parker, N.A. (1998): Mechanics of small utility cuts in urban street pavements: implications for restoration. Journal of the Transportation Research Record, No. 1629, pp. 226-234.

Hunter, A. (2005): Effect of Trenchless Technologies on Existing Iron Pipelines. Proceedings of The Institution of Civil Engineers- Geotechnical Engineering, Vol. 158, No. 3, pp. 159-167.

Irigoyen, J. L. (2003): Millennium. Development. Goals:The infrastructure contribution. World Bank.[Online]http://www.worldbank.org/transport/forum2003/presentations/ irigoyen.ppt [Access date: 10th September, 2007].

Iseley, T. and Gokhale, S. (1997): Trenchless installation of conduits beneath roadways. Synthesis of Highway Practice 242. Washington DC: Transportation Research Board.

Iseley, T. and Tanwani, R. (1990): Social cost of traditional methods of utility installation: Proceeding of first Trenchless Excavation Symposium, Houston, Texas. AA1-AA7.

Jean-Philippe Montillet, Ahmad Taha, Xiaolin Meng, Gethin W. Roberts (2007): Mapping the Underworld: Testing GPS and GSM in Urban Canyons; GPS World Mar 1, 2007

Jensen, K. A, Vernon R. Schaefer, Muhannad T. Suleiman and David J. White (2005): Characterization of Utility Cut Pavement Settlement and Repair; Techniques Proceedings of the 2005 Mid-Continent Transportation Research Symposium, Ames, Iowa.

Jesdanun, A. (2004): "GE Energy acknowledges blackout bug," The Associated Press, http://www.securityfocus.com/news/8032. Accessed April 7, 2004.

Jongensen, R. & Associates (1987): Cost and safety effectiveness of highway design elements. Report 197. Washington: National Cooperative Highway Research program.

Joseph, J. (2004): "Energy Use in the Municipal Water/Wastewater Treatment Sector," Presentation at the New York Regional Energy Workshop, New York, NY: Columbia University.

Joshua, S. C. and Garber, N. J. (1990): Estimating truck accident rate and involvements using linear and Poisson regression models. Transportation Planning and technology 15:41-58.

Joungro-Ku, S. (2002): National Police Agency Reasons for Declining Road Traffic Accidents and Future Plans: Interim Report, Korea National Police Agency.

Jovanis, P. P. and Chang, H. L. (1986): Modelling the relationship of accidents to mile travelled.

Transportation Research Record 1068:42-51.

Khattak, A. J. (2006): Intersection safety. Transportation studies: NDOR Research Project No. SPR-1(2) P544-SJ0105.

Khigali, W.E.I. and El Hussein, M. 1999. Managing utility cuts: Issues and considerations. NCRR/CPWA Seminar Series: Innovations in Urban Infrastructure. Washington, DC: APWA International Public Works Congress

Knight, M.A. Tighe, S.T. and Adedapo, A. (2004): Trenchless Installations Preserve Pavement Integrity. Proceedings Annual Conference of the Transportation Association of Canada, Quebec City, September 2004.

Knight, M. Duyvestyn, G. and Gelinas, M. (2001): Excavation of surface installed pipeline. Journal of Infrastructure Systems, Vol.7, No3.

KNWD (2006): Kenya Water Report 2005: Water for wealth creation and healthy environment for a working nation. Kenya National Water Development [Online] http://www.irc.nl/page/32007 [Access date: 20th September 2007]

Kraay, J. H. and Leidschendam, D. A. (1989): Safety aspects of Urban Infrastructure. SWOV Institute for Road Safety Research, The Netherlands, R-89-14.

Kuhn, B. Brydia, R. Jasek, D. Parham, A. Ullman, B. and Blaschke, B. (2003): Utility Corridor Structures and Other Utility Accommodation Alternatives in TxDOT Right-of-Way Project Summary Report 4149-1, Texas Transportation Institute, Texas A&M.

Kuwano, R. (2006): Hidden Cavities Under the ground – Their Causes and Consequences, Icus Newsletter, International Center for Urban Safety Engineering, Institute of Industrial Science, The University of Tokyo, Volume 6 Number 2, July-September 2006.

Lam, D. Ang, T.-C. and Chiew, S.-P. (2003): Structural Steelwork: Design to Limit State Theory, Butterworth-Heinemann, 3rd.

Lee, J. and Mannering, F. (2002): Impact of roadside features on the frequency and severity of run-off-roadway accidents: an empirical analysis. Accident Analysis and Prevention 34:149-161.

Lee, K. H. (2005): First Course on Fuzzy Theory and Applications, Springer- Verlag Berlin Heidelberg.

Lee, S.Q.S. and Lauter, K.A. (1999): Impact of Utility Trenching and APpurtenances on Pavement Performance in Ottawa-Carleton, report prepared by Trow Consulting Engineers, Ltd for Environment and Transportation Department, Regional Municipality of Ottawa-Carleton, Ottawa, Ontario, Canada, July 29, revised August 13.

Lee, S.Q.S. and Lauter, K.A. (2000): Using pavement management system concepts to determine the cost and impact of utility trenching on an urban road network Transportation Research Record, no. 1699, pp. 33-41

Legge, J. S.Jr. (1991): Traffic Safety Reform in the United States and Great Britain. University of Pittsburg Press.

Leipziger, D. (2003): Millennium. Development. Goals:The infrastructure contribution. World Bank.   [Online] http://www1.worldbank.org/prem/prmpo/povertyday/docs/2003/leipziger.pdf. [Access date: 10th September, 2007]

Lippert, A. (2004): "Critical Infrastructure Protection for the Energy Sector," 2004 Asia-Pacific

Homeland Security Summit Briefing

Lundqvist, L. Mattsson, L. and Kim, T. J. (1998): Network Infrastructure and the Urban Environment. Advances in Spatial Science.

Luo, M. (2004): "The Next Stop for the Subway Is a Fully Automated Future," New York Times, p. B1.

Maher, M. J. and Summersgill, J. (1996): A comprehensive methodology for the fitting of predictive accident models. Accident Analysis And Prevention. 28(3):281-296.

Majumdar, K. K. and Majumder, D. D. (2004): Some Studies on Uncertainity Management in Dynamical Systems using Cybernetic Approaches and Fuzzy Techniques with Applications. International Journal of Systems Science, 35(15), pp. 889-901(13).

Mangolds, A. and Carapezza, J. (1991): Assessment of Pavement Cutback Requirements, Foster-Miller, Inc. Final Report, RD & D Project D.28.3, report BUG-D283-FM- 9081-429 prepared for Brooklyn Union Gas Company, Research Development and Demonstration Department, Brooklyn, NY and Consolidated Edison Co. of New York, Inc. Research and Development, New York, NY.

Mannering, F. and Kilareski, W. (1998): Principles of Highway Engineering and Traffic Analysis. New York: John Wiley & Sons, NY (2nd Edition), pp 340.

Masada, T. (2000): Modified Iowa formula for vertical deflection of buried flexible pipe. Journal of Transportation Engineering, ASCE, Vol. 126, No. 5.

Maycock, G. (1996): Generalized Linear Models in the Analysis of Road Accidents. Transport and Road Research Laboratory.

McKenzie, D. (2005): "Review of Issues Affecting Utilities and Road, Rail and Motorway Corridors" (2005). Submission to Ministry of Economic Development – IPENZ (Institution of Professional Engineers NZ) Transportation Group August 2005.

Melvin, E. (1992): Plan, Predict, Prevent: How to Reinvest in Public Buildings, American Public Works Assocation Research Foundation, Chicago.

Miaou, S. P. and Lum, H. (1993): "Modelling Vehicle Accidents and Highway Geometric Design Relations. Accident Analysis and Prevention 24(6), pp. 689-709.

Miaou, S. P. (1994): The relationship between Truck Accidents and Geometric Design of Road Sections: Poisson versus Negative Binomial Regressions. Accident Analysis and Prevention 26(4): 471-482.

Miaou, S. P.; Hu, P. S.; Wright, T.; Davis, S. C. and Rathi, A. K. (1991): Development of relationships between truck accidents and Highway Geometric Design: Phase 1. Technical memorandum prepared by the Oak Ridge National Laboratory. Washington, DC: federal Highway Administration.

Millennium-Project (2005): Investing in Development: A Practical Plan to Achieve the Millenium Development Goals. Overview. United Nations Millennium Project. [Online] http://www.unmillenniumproject.org/reports/index.htm [Access date: 13th September, 2007]

Milton, J. and Mannering, F. (1998): The relationship among highway geometrics, traffic-related elements and motor-vehicle accident frequencies. Transportation 25: 395-413.

MoRPW (2000): Road Maintenance Manual Part - 1. Government of the Republic of Kenya, Ministry of Roads and Public Works (MoRPW).

MoTC (2004): Sub-sector Policy Papers and Implementation Matrices. Ministry of Transport & Communications, Government of the Republic of Kenya. [Online] http://www.krb.go.ke/Transportpolicy.php/ [Access date: 13th September, 2007]

Moteff, J. D. (2000): Critical Infrastructures: Background and Early Implementation of PDD-63. A Report of the Congressional Research Service (RL30153).

Motro, A. and Smets, P. (1997): Uncertainty Management in Information System: From Needs to Solutions, Kulwer Academic Publishers, Boston/ London/ Dordrecht.

Mulvey, A. (2002): Gender, Economic Context, Perceptions of Safety, and Quality of Lif. A Case Study of Lowell, Massachusetts (U.S.A.), 1982–96 in American Journal of Community Psychology, October 2002. 30 (5): 655-679(25) Kluwer Academic Publishers.

NACUBO (1994): Toward Infrastructure Improvement: An Agenda for Research, Building Research Board, National Research Council. Building Research Board, National Research Council, National Academy Press, Washington, DC, 129.

Najm, H. (2005): Study of the Effects of Buried Pipe Integrity on Roadway Subsidence, Final report, Pipe-RU6558 In cooperation with Oldcastle Pipe Company And U.S. Department of Transportation; Federal Highway Administration.

National Association of Regulatory Utility Commissioners (2004): Resolution on Utility Sector Interdependencies.

National Bureau of Statistics (2002): Population and Housing Census. National Bureau of Statistics, Tanzania.

National Research Council (2002): Making the Nation Safer, Washington, DC: National Academy Press.

North American Electric Reliability Council (2004): Gas/Electricity Interdependencies and Recommendations. Gas/Electricity Interdependency Task Force, NERC Planning Committee.

Nozick, L. K. and Turnquist, M. A. (2005): Assessing the Performance of Interdependent Infrastructures and Optimizing Investment. International Journal of Critical Infrastructures, 1(2/3), pp.144-154.

NRC (2002): Down to Earth: Geographic Information for Sustainable Development in Africa, National Research Council, National Academies Press

NRC (2006): A Framework for Municipal Infrastructure Management for Canadian Municipalities. National Research Council. [Online] http://irc.nrc-cnrc.gc.ca/pubs/fulltext/b5123.7/ [Access date: 13th September 2007]

NTP-Committee (2004): Recommendations on Integrated National Transport Policy : Moving a Working Nation. National Transport Policy Committee, Ministry of Transport & Communications, Government of the Republic of Kenya. [Online] http://www.krb.go.ke/Transportpolicy.php [Access date: 13th September, 2007]

ODA (1991): Towards Safer Roads in Developing Countries: A guide for Planners and Engineers. ODA, Berkshire.

Ogal, N. A. (2008): Toward Development of a prototype tool for integrated Infrastructure asset management activities: Characterization and performance measurement of the urban infrastructure system MSc Thesis (MWI –2008/27) UNESCO-IHE Institute for

Water Education.

Ogden, K. W. (1992): Urban Goods Movement: A Guide to Policy and Planning.

Ogden, K.W. (1996): Safer Roads: A Guide to Road Safety Engineering, Ashgate Publishing Limited, U.K.

Ojovan, M. and Lee, W. (2006): Topologically Disordered Systems at the Glass Transition. Journal of Physics, 18(50), 11507 – 11520.

Olowu, D. and Wunsch, J. S. (2003): Local Governance in Africa: the challenges of democratic decentralization Lynne Rienner.

Payne C. (2002): New York's Forgotten Substations. The Power Behind the Subway, New York, NY: Princeton Architectural Press.

Peerenboom, J. P. (2001): "Infrastructure Interdependencies: Overview of Concepts and Terminology," Infrastructure Assurance Center, Argonne National Laboratory, Argonne IL 60439.

Peerenboom, J. P., Fisher, R. E., Rinaldi, S. M. and Kelly, T. K. (2002): Studying the Chain Reaction, Electric Perspectives (EEI).

Peerenboom, J. P., Fisher, R. E. and Whitfield, R. (2001): "Recovering from Disruptions of Interdependent Critical Infrastructures," prepared for CRIS/DRM/IIIT/NSF Workshop on "Mitigating the Vulnerability of Critical Infrastructures to Catastrophic Failures" Lyceum, Alexandria, Virginia.

Peters, T. (2002): City combats damage to city streets caused by utility cuts. Public Works 133 (4).

Rafferty, M. (2007): Reductionism, Holism and Systems Thinking. London South Bank University.   [Online]
http://www.systemdynamics.org/conterences/2007/proceed/paper/RAFFF116
[Access date: 22nd February, 2008].

Rajani, B and Makar, J. A. (2000): Methodology to estimate remaining service life of grey cast iron water mains. Canadian Journal Civil Engineering, 2000, 27, 1259-1272.

Rajani, B. and Tesfamariam, S. (2004): Uncoupled axial, flexural, and circumferential pipe-soil interaction analyses of jointed water mains. Canadian Geotechnical Journal, 2004, 41, 997-1010.

Rajani, B., Zhan, C. and Kuraoka, S. (1996): Pipe-soil interaction analysis for jointed water mains. Canadian Geotechnical Journal, 1996, 33, 393-404.

Réka, A., Hawoong, J. and Albert László, B. (2000): Error and Attack Tolerance of Complex Networks. International WeeklyJournal of Science(406), 378-382.

Rinaldi, S. M., Peerenboom, J. P. and Kelly, T. K. (2004): Identifying, Understanding, and Analyzing Critical Infrastructure Interdependencies - IEEE Control Systems Magazine. [Online] http://www.ce.cmu.edu/~hsm/im2004/readings/CII-Rinaldi.pdf [Access date: 20th September 2007]

Rinaldi, S. M., Peerenboom, J. P. and Kelly T. K. (2001): "Identifying, Understanding and Analyzing Critical Infrastructure Interdependencies," IEEE Control Systems magazine.

Robinson, P. Woodard J. B. and Varnado, S. G. (1998): "Critical Infrastructure: Interlinked and Vulnerable," Issues in Science and Technology Online.

Ross, T. J. (1995): Fuzzy Logic with Engineering Application, Mc Graw-Hill, New York.

Ruyters, H. G. Slop M. and Wegman, F.C. (1994): Safety Effects of Road Design Standards; SWOV Institute for road safety Research, Leidschendam.

Selig, E.T. (1988): Soil Parameters for Design of Buried Pipelines, Pipeline Infrastructure. Proceedings of the Conference, American Society of Civil Engineers, New York, NY, pp. 99-116.

Singh, K. Steinberg, F. and Von Einsiedel, N. (1996): Integrated Urban Structure Development in Asia.

Schaefer, V., Suleiman, M., White, D., Swan, C. and Jensen, K. (2005): Utility cut repair techniques—investigation of improved cut repair techniques to reduce settlement in repaired areas. Iowa Highway Research Board. Iowa Department of Transportation, 800 Lincoln Way Ames, IA 50010. Report # IHRB Project TR-503.

Schandersson, R (1994): Road pavement condition and traffic safety: Swedish Road and Transport Research Institute VTI.

Scheirs, J. (2000): Compositional and Failure Analysis of Polymers - A Practical Approach, John Wileys & Sons, LTD, New York.

Schmandt, H. (1996): the quality of urban life, Beverly Hills, London.

Schouten, M. A. C. (2007): Project Management. Unpublished Lecture notes UNESCO-IHE (LNO325/06/1), Delft, The Netherlands.

Schubeler, P. (1996): Participation and Partinership in Urban Management Program, The World Bank, Washington, D.C, USA.

Shahin, M. Y. and Kohn, S. D. (1981): Pavement maintenance management for roads and parking Lots. US Army Corps of Engineers, Construction Engineering Research Laboratory (CERL), Technical Report M-294

Shahin, M. Y. Crovetti, J. A. and J. L. Franco, (1986): Effects of Utility Cut Patching on Pavement Performance and Rehabilitation Costs, Annual Meeting of the Transportation Research Board, Washington, D. C. December 23.

Shahin, M.Y. and Crovetti J A. (1987): Determining the effect of Utility Cut patching on service life prediction of Asphalt concrete pavement, 2nd North American conference on managing pavements, proceedings Vol. 1, Toronto, Canada, pp. 225-236.

Shield, R.T. (1955): On Coulomb's law of failure in soils. Journal of the mechanics and physics of solids, Vol. 4, Issue 1, pp. 10-16.

Shook, J. F. Finn, F. N. Witczak, M. W. and Monismith, C. L. (1982): Thickness Design of Asphalt Pavements - the Asphalt Institute Method. 5th International Conference on the Structural Design of Asphalt Pavement, pp. 17-44.

Smith, W.H. (1976): Frost loadings on underground pipe. Journal of the American Water; Works Association, Vol. 68, No. 12, pp. 673–674.

Smutzer, R. Brunner, T Scheidler B. and Downey Kent D. (2004): "Accountability, Communication, Coordination and Cooperation" Report of the Utility Relocation Task force, Wisconsin Department of Transportation

SUDAS (2004): Urban Standard Specifications for Public Improvements Manual. Urban Standard Specifications Committee, Central Iowa Metropolitan Areas and Municipalities.

SUDP (2002): Toward a more Sustainable City Planning, Department of social and Economic geography, Umea University.

Tanzania National Roads Agency (2008): A Study on Management and Preservation of Road Reserve along Trunk and Regional Roads. Final Report.

Tarakji, G. (1995): Effect of Utility Cuts on the Service Life of Pavements in San Francisco, Study Procedure and Findings, Final report by Engineering Design Center, San Francisco State University for the Department of Public Works, City and Country of San Francisco, San Francisco, CA.

Tesfamariam, S. Rajani, B. and Sadiq, R. (2006): Possibilistic Approach for Consideration of Uncertainties to Estimate Structural Capacity of Ageing Cast Iron Water Mains. Canadian Journal of Civil Engineering.

The 9-11 Commission (2004): Final Report of the National Commission on Terrorist Attacks Upon the United States. Washington, DC: the 9/11 Commission

Thissen, W.A. H. and P. M. Herder (2003): Critical Infrastructures" State of the Art in Research and Application. (Boston: Kluwer Academic Publishers).

Thomas, W.H. M.J. North, C.M. Macal, and J.P. Peerenboom. "From Physics to Finances: Complex Adaptive Systems Representation of Infrastructure Interdependencies," Naval Surface Warfare Center Technical Digest, Naval Surface Warfare Center, Dahlgren, VA USA (In Press).

Tighe, S. Jeffray, A. Kennepohl, G. Haas, R. and Matheson, M. (2003): Field experiments in CPATT's long-term program of pavement research. Proceedings Transportation Association of Canada Annual Conference, St. John's, NF.

Tighe. S. Knight, M. Papoutsis, D. Rodriguez, V. and Walker, C. (2002): User Cost Savings in eliminating pavement excavation through employing trenchless technology. Canadian Journal of Civil Engineering Vol. 29, pp.751-761.

Tighe. S. Lee, T. McKim, R. and Hass, R. (1999): Traffic delay cost savings associated with trenchless technology. Journal of Infrastructure Systems. Vol. 5, No. 2, pp.45-51.

Titus-Glover, L. (1995): Evaluation of Pavement Base and Subgrade Material Properties and Test Procedures. MSc. Thesis. Texas A &M University, College Station, Texas.

Todres, H. A. and C. L. Wu, (1990): Theoretical Analysis of Utility Cut Restoration in Asphaltic Pavement, Vol.ume I, report by Construction Technologies Laboratories, Inc. submitted to Institute of Gas Technology, Chicago, IL, October.

Todres, H. and Saha N. (1996): Asphalt Paving repairs Study – Theoretical Modeling, APWA Reporter November 1996.

Todres, H. A. (1999): Utility Cuts in Asphalt Pavements: Best Engineering Practices, in Innovations in Urban Infrastructure, APWA International Public Works Congress and Exhibition, Las Vegas, Nevada, September 14-17.

Todres, H.A. and P.E. Baker (1996): Utility Research in Pavement Restoration, Presented at APWA International Public Works Congress and Exposition, Washington, D.C.

Tolliver, D. (1994): Highway Impact Assessment. Techniques and Procedures for Transportation Planners and Managers.

Transportation Association of Canada (2008): Management of utilities in and adjacent to the public right-of-way: Survey of Practices August 2008.

Transportation Research Board. (1987): Design safer roads. Special report 214. Washington, DC: National Research council: 1987.

Tromp, J. P. M. (1987): Ruts and waves in the road surface. Institute for Road Safety Research SWOV, pp.124-132.

TRRL, (1991): Towards Safer Road in the developing countries. Ross Silcock Partnership, Newcastle Upon Tyne, England.

Turner. D and Mansfiel. E (1990): Urban Trees and Roadside Safety. Journal of Transportation Engineering, Vol. 116, No 1, pp. 90-104.

Ullman, B. and Blaschke, B. (2003): Utility Corridor Structures and Other Utility Accommodation Alternatives in TxDOT Right-of-Way Project Summary Report 4149-1, Texas Transportation Institute, Texas A&M.

Ugwu, O. O. and Haupt, T. C. (2007): Key performance indicators and assessment methods for infrastructure sustainability - a South African construction industry perspective. Building and Environment, 42(2), 665 - 680.

U.S. EPA (2000): Development, Community and Environment - Our Built and Natural Environments, Washington, DC: U.S. EPA.

U.S. Executive Office of the President (1998): "The Clinton Administration's Policy on Critical Infrastructure Protection: Presidential Decision Directive 63".

U.S.-Canada Power System Outage Task Force (2003): Final Report; Blackout in the United States and Canada: Causes and Recommendations. The Task Force, April, 2004.

UNDESA (2004): Millennium Development Goals: Progress Report. Statistics Division, United Nations Department of Economic and Social Affairs (UNDESA). [Online] http://www.un.org/milleniumgoals/mdg2004chart [Access date: 1st April, 2008]

UNDESA (2007): The Millenium Development Goals Report 2007. Statistics Division, United Nations Department of Economic and Social Affairs (UNDESA), New York.

UNDG (2003): Indicators for Monitoring the Millennium Development Goals. United Nations Development Group. [Online] http://www.development goals.com/UNDG%20document_final [Access date: 1st April, 2008]

UNDG (2005): The Millennium Declaration and the MDGs. United Nations Development Group. [Online] http://www.undg.org/content.cfm?id=502 [Access date: 13th September, 2007]

Uni-Bell (1993): Handbook of PVC Pipes, 3rd Edition published by Uni-Bell PVC Pipe Association, Dallas, TX.

Vallerga, N. & Associates, 2000. Final Report: Impact of Utility Cuts on Performance of Seattle Streets, report submitted to City of Seattle, Seattle Transportation, project 178.01.30, January 31.

Van der Hoff, R. and Steinberg, F. (1992): Innovative Approaches to Urban Development. Institute for Housing and Urban Development Studies Rotterdam iHs.

Van Dijk, M. P. (2006): Managing Cities in Developing Countries: The Theory and Practice of Urban Management. Published by Edward Elgar publishing Limited, Glos GL50 IUA, UK.

Vogt, A. and Bared, J. G. (1998): Accident Models for Two-lane Rural Roads: Segments and Intersections. SWOV Publication No. FHWA-RD-98-133.

WalaaB, E. I. and Elhussein H. MOHAMED (1999): Managing utility cuts issues and considerations APWA International Public Works Congress NRCC/CPWA Seminar

Series "Innovations in Urban Infrastructure".

Watkins, R. K. (1966): Structural Design of Buried Circular Conduits. Highway Research Board. No 145, pp. 1-17.

Watkins, R. K. and Anderson, L.R (2000): Structural mechanics of buried pipes. CRC. Press, New York.

WEFA (2007): The Africa Competitiveness Report, 2007. World Economic Forum for Africa. [Online] http://www.weforum.org/pdf/gcr/africa/kenya.pdf [Access date: 14th September]

Wegman, Fred (1996): Road design, Human behaviour, and road accidents: towards a learning design community. Institute for Road Safety Research (SWOV) Leidschendam, 1996.

Weir, C. H. and Classen J. (1986): "The Corridor Concept-Theory and Application" Right of Way August 1986.

West, D. B. (2001): Introduction to Graph Theory, Prentice Hall, 2nd Edition.

White, T. D. (1985): Pavement Instrumentation: guide to instrumentation plan for flexible and rigid pavements. FHWA/EP-621-01, Demonstrations Projects Division, FHWA Washington, DC.

WHO (2000): Setting New Directions in Traffic safety. Traffic safety Center, Geneva.

Wilbur, S. and OOP (1993): Kenya Municipal Study: Road Condition survey for Municipality of Kisumu. Government of the Republic of Kenya, Ministry of Local Government, Department of Urban Development, Kisumu.

Williams, R. (1996): Longitudinal Occupancy of Controlled Access Right-of-Way by Utilities, NCHRP Synthesis 224, Transportation Research Board, NRC, 1996, Washington, DC;

Wood, G. R. (2001): Generalized linear accident models and goodness of fit testing

World Bank, (1994): Infrastructure for Development: World Development Report 1994, Oxford University Press.

World Bank, Road Safety: http://www.worldbank.org/transport/roads/safety.htm#crash

Zimmerman, R. (2001): "Social Implications of Infrastructure Network Interactions," Journal of Urban Technology, Vol. 8, No. 3.

Zimmerman R. (2003): "Public Infrastructure Service Flexibility for Response and Recovery in the September 11th, 2001 Attacks at the World Trade Center," in Natural Hazards Research & Applications Information Center, Public Entity Risk Institute, and Institute for Civil Infrastructure Systems, Beyond September 11th: An Account of Post-Disaster Research, Special Publication; #39. Boulder, CO: University of Colorado.

Zimmerman R .and T.A. Horan, eds. (2004): Digital Infrastructures: Enabling Civil and Environmental Systems through Information Technology, London, UK: Routledge.

Zimmerman, R (2006): Decision-making and the Vulnerability of Interdependent Critical Infrastructure Report by Institute for Civil Infrastructure Systems (ICIS); Robert F. Wagner Graduate School of Public Service New York University.

Annexes

Annex I

A. Questionnaires for road users in Dar es Salaam road corridors

---

Background information

1. Date of the interview (dd/mm/yyyy) ___/___/_____
2. Day of the interview (Monday to Friday)........................
3. Name of the interviewer.................................
4. Interview team ....
5. Specific road/locations (see the codes below)
   1. Morogoro road
   2. Kilwa road
   3. Bagamoyo road
   4. Nyerere road
   5. Mandela road

Characteristics of the road users

6. Sex of the respondent
          1. Male   2. Female
7. Age of the respondent
   1. Less than 18 years
   2. Between 18 and 25 years
   3. Between 25 and 59 years
   4. 60 years and above
8. Level of education of the respondent (see the table below)

No formal Schooling	Adult literacy	Primary Education 1 – 4	Primary education 1 - 7	Secondary education	Vocational training	College diploma	University degree
1	2	3	5	6	7	8	9

9. Respondent's main occupation (see the codes below)
   1. Full time employee
   2. Driver
   3. Business/petty trade
   4. Student
   5. Farmer
   6. Other (state)
10 How frequently do you use the following means of transport when using the roads under this study (tick where is appropriate)

Transport means	always	frequently	rare
a. Commuter Bus/daladala			
b. Private Vehicle			

c. Motor cycle				
d. Bicycle				
e. Walk				
f. Any other means (please state)......................				

11. What are your daily common trips
   a.   Travelling to the work place
   b.   Travelling to the market
   c.   Travelling to school/college
   d.   Travelling to visit friends/relatives
   e.   Driving people from one place to another (if a driver)
   f.   Travelling to any other place (specify)

12. Which road do you commonly use in your trips
   a.   Morogoro road
   b.   Kilwa road
   c.   Bagamoyo road
   d.   Nyerere road
   e.   Mandela road
   f.   Any other (please mention)

13. Which among the following are periods of high traffic on the roads you commonly use

06hrs -09hrs	09hrs-12hrs	12hrs-15hrs	15hrs-18hrs	18hrs-21hrs

Characteristics of the interaction of the infrastructure assets in urban road corridor in relation with movement and safety

14. Which of the following infrastructure exist in the road you use

1	Water Pipes	
2	Sewer Pipes	
3	Telephone Cables	
4	Electric Cables	

15. For the following mentioned infrastructure how did you come to know of their existence along the roads you use

	Seeing overhead cables	Seeing manholes	Seeing leakages	Seeing open trenches	Seeing mark post
Water Pipes					
Sewer Pipes					
Telephone Cables					
Electric Cables					

16. How frequently is your movement being disturbed with condition of the following infrastructure elements that exist within the roads you commonly use.

	Always	Several times	Occasionally	Seldom
Road Surface conditions e.g. skid resistance, water layer, corrugations etc	1	2	3	4

Burst of water supply pipes	1	2	3	4
Burst of sewage systems pipes	1	2	3	4
Cutting the road for the crossing underground infrastructure	1	2	3	4
Leakages of water pipes/sewages on road surfaces	1	2	3	4
Pot holes on the road surface	1	2	3	4
Sewage openings on the road surfaces	1	2	3	4
Railway crossing	1	2	3	4
Gas pipes	1	2	3	4
Road marking	1	2	3	4
Settlement of the road due to failure of underground Infrastructure	1	2	3	4
Open manholes on the road surface	1	2	3	4

17. Which of the seasons of the year would you think road movements are mostly unsafe?
   1. Dry seasons   2. Wet/rainy seasons   3. Both dry and wet seasons
   Please give reasons for your answer in question the above.............

18. Which of the following roadside infrastructure items do you think frequently cause unsafe movement of people in the public right of way (Note that one can pick more than one items)

Features	Always	Several Times	Occasionally	Seldom
Guard Rail				
Embankment				
Un clear sight distance				
Road Signs				
Utility poles				
Drainage systems				
Pot holes on the shoulder				
Manholes on the shoulders				
Traffic lights				
Street lights				
Advertisement posts				

What do you think are the effect of the above listed to the movement and safety? And how do you feel when you find these items on the road.

19. When are the road corridor movements likely to be unsafe regarding the construction, maintenance and break down of infrastructures? (put a tick (√)in the cell of the choice)

	Water pipes	Utility poles	Gas pipes	Roads	Underground cables	Sewages	Overhead cables
Construction							
Maintenance							
Break down of the system							

   Explain how maintenance and construction of the above mentioned activities cause unsafe movement in the road corridor?
   Maintenance

....

Construction

....

Repairing the system

Explain how infrastructure in the table is affecting movement and safety of movement

Effects of the land use, population and travel characteristics on movement and safety movements in the road corridor

20. i) Which of the following features could you list as the characteristic features of the public road corridor of the road that you commonly use? (Note that one can pick more than one items)

      a. Market places along the roads, with people spreading their commodities on the road surfaces

      b. Bars built in the road corridor/ or people sitting along the road for a drink

      c. Unofficial or random parking of cars or trucks along the roads in the road corridor

      d. Vendors' huts erected on the public right of way

      e. Vendors who are always moving along the roads selling goods

      f. Shopping centres

      g. Office premises

      h. Residential areas

      i. Bus buys

      j. Parking places

      k. Schools

      l. Hotel and restaurant

      m. Any other, Mention

ii) Mark on the above land use which has influence to unsafe movements in the public road corridor?

Please explain how each of the above contribute to unsafe movements in the public right-of-way?

....
....

21. What category/categories of road users do you think they mostly experience unsafe movements in the road corridor? See the codes below

      a. Pedestrians

      b. People who travel by public means/daladala

      c. People who use private means of transport

      d. People who use motorcycles

      e. People who use bicycles

      f. Tricycle

      g. Cart haulers

Please explain your answer

....

Infrastructure designs, construction and maintenance problems existing in the public road corridor

22. i) Do you think there are infrastructures planning and design problems in the public road corridor that you commonly use? .... (Yes/No)

If the answer is in above question is yes; which of the following are those problems? If any other mention

	Not at all	Rarely	Often	Always
Incompatibility among infrastructure facilities				
Poor choice of materials				

Lack of integrated planning				
Lack of adequate resources				
Unregulated land use				

How the above-mentioned items do result into unsafe movement in the public road corridor?

iii) To what extent do operational and maintenance activities in the following facilities affect movement and safety in the public road corridor?

	Not at all	Rarely	Often	Always
Utility Poles				
Gas pipes				
Sewages				
Water Pipe				
Roads				
Cables				
Advertisement				
Concrete pole				
Traffic Light				
Street Light				
Drainages				

(iv) To what extent do operational and maintenance activities in the following facilities affects movement in the public road corridor?

	Not at all	Rarely	Often	Always
Utility Poles				
Gas pipes				
Sewages				
Water Pipe				
Roads				
Cables				
Advertisement				
Concrete pole				
Traffic Light				
Street Light				
Drainages				

How do the above-mentioned items cause unsafe and difficulties in mobility in the public road corridor?

S/N		Frequently	Rare	None
1	Infection Disease			
2	Banditry			
3	Other accidents			
4	Fear of being involved in an accident			
5	Road traffic accident			
6	Constriction of spaces			

B. Survey Forms

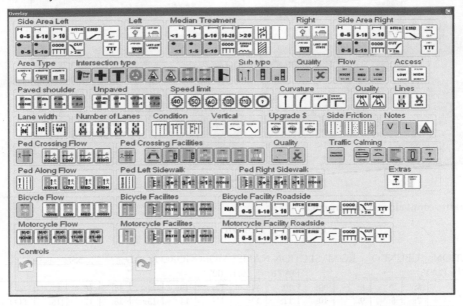

C. Set up of the survellance cameras

 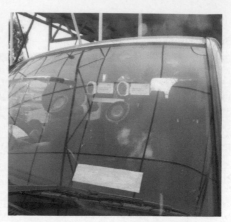

D. Traffic count along Morogoro road

FORM FOR TRAFFIC COUNTS

LOCATION:__UBUNGO___ROAD SECTION NAME:_MOROGORO ROAD_____ SS:_____

NIGHT / DAY:_____   DATE:___19/1/2007_____

TIME	6: 7:	7: 8:	8: 9:	09: 10:	10: 11:	11: 12:	12: 13:	13: 14:	14: 15:	15: 16:	16: 17:	17: 18:	TOTAL
Pedestrian	1122	690	473	310	277	212	191	213	240	180	228	286	4422
Other Non-Motorized	30	30	35	49	94	38	79	45	48	66	117	72	703
Motor cycle	42	69	39	51	68	104	69	46	34	61	113	116	812
Cars	860	803	839	725	397	378	316	320	413	385	444	468	6348
P/Ups & Vans	54	99	109	43	82	43	63	89	81	118	135	182	1098
Light veh	23	50	35	24	25	30	41	31	26	62	28	35	420
Medium veh	19	14	13	17	5	8	18	10	7	22	15	20	168
Heavy veh	9	9	6	9	-	-	-	-	-	25	11	20	89
Very heavy veh	12	5	5	-	-	-	-	-	3	19	17	20	81
Buses Under 25 Seats	270	372	401	439	374	283	291	296	330	432	446	572	4506
Buses Over 25 Seats	142	166	138	220	216	125	156	161	189	163	238	278	2192
Other Vehicles	59	52	57	80	33	54	53	22	65	84	104	41	704
TOTAL	2642	2359	1950	1967	1571	1275	1277	1233	1436	1617	1896	2110	2133

Annex II

A. Road Traffic Accident

A.1 Road traffic accidents along the corridors

YEAR	2000	2001	2002	2003	2004	2005	2006	2007	TOTAL	%
BAGAMOYO ROAD	1,006	953	973	935	1147	1167	1390	1586	**9,157**	26
KILWA ROAD	477	503	497	528	663	649	677	731	**4,725**	13
MANDELA ROAD	451	491	417	517	549	567	593	608	**4,193**	12
MOROGORO ROAD	1,416	1,605	1776	1757	1598	1853	1913	1939	**13,857**	39
NYERERE ROAD	471	421	401	415	434	485	501	559	**3,687**	10
TOTAL	**3,821**	**3,973**	**4,064**	**4,152**	**4,391**	**4,721**	**5,074**	**5,423**	**35,619**	**100**

A.2 Road traffic accidents along the links on the study sites

ROAD NAME	LINK SITE	NUMBER OF ACCIDENTS ON THE JUNCTION SITES							
		2000	2001	2002	2003	2004	2005	2006	2007
BAGAMOYO ROAD	Tanganyika Motors	9	6	0	1	4	4	5	0
	Aga Khan	0	0	0	1	0	4	2	0
	Salender Bridge	35	90	33	3	5	18	16	16
	Kinondoni junction	0	0	0	0	0	0	0	0
	St. Peters	6	9	30	0	0	8	7	7
	Namanga	0	16	32	0	14	8	19	32
	Morocco	19	15	31	11	20	18	12	42
	Victoria	44	22	95	24	12	26	16	73
	Sayansi	50	20	91	6	16	16	10	23
	Bamaga	0	6	29	8	0	8	7	47
	Mwenge	42	36	63	1	16	34	35	33
MANDELA ROAD	Bandari	0	0	0	0	0	0	0	0
	Kurasini	0	0	0	0	0	0	0	0
	Uhasibu	0	0	0	0	0	0	0	0
	Uwanja wa Taifa	0	0	0	0	0	0	0	0
	Chang'ombe Polisi	5	3	2	1	1	0	0	5
	Temeke	0	0	0	0	0	0	0	0
	Veternary	11	6	4	4	50	14	0	30
	Tazara	37	93	62	61	16	0	0	54
	Buguruni	37	46	8	8	30	17	21	35
	Tabata	0	15	27	21	14	1	88	52
	External	0	43	15	22	64	23	74	27
	Ubungo	0	0	0	0	0	0	0	0
	Mlimani city	0	6	3	2	1	0	7	5
	Mwenge Roundabout	0	0	0	0	0	0	0	0
	Mwenge	0	0	0	0	0	14	0	0
MOROGORO ROAD	Samora Avenue	0	0	0	0	0	0	0	0
	Akiba	0	0	0	11	0	23	20	0
	Tanzania Region	0	0	0	0	0	0	0	0
	Lumumba	0	0	0	0	0	0	0	0
	Msimbazi	0	0	0	0	0	0	0	0
	Fire	8	0	1	14	6	15	16	0
	Magomeni	86	51	177	64	14	29	23	128
	Kagera	11	1	0	11	0	0	0	0
	Urafiki	0	0	0	0	0	0	0	0
	Shekilango	0	10	0	6	2	17	0	0
	Ubungo	9	15	5	17	75	18	28	0

B. Traffic Volume

B.1 Peak hour volume along Morogoro

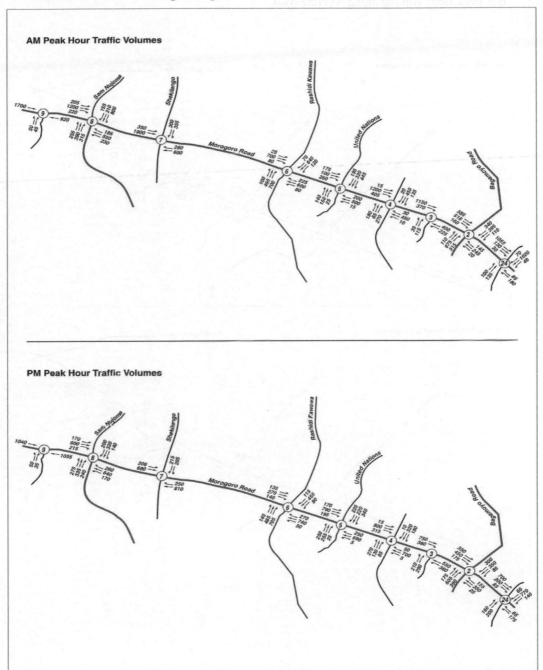

B.2 Peak hour volume along Nyerere road

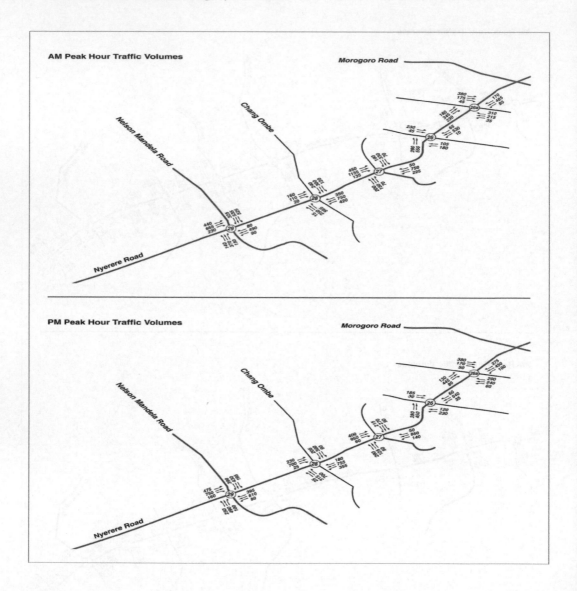

C. Road corridor Inventory survey
C.1 Road way roughness data along Morogoro road

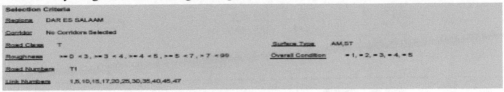

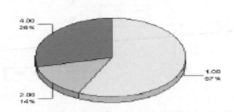

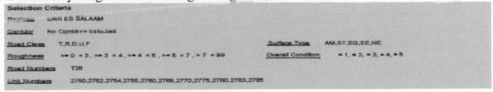

	EIRI (m/km)					
	1	**2**	**3**	**4**	**5**	Total number of sub-links
	< 3	3 to < 4	4 to < 5	5 to < 7	>= 7	
No. of Sublinks	4	1	0	2	0	7
% of Sublinks	57.14	14.29	0.00	28.57	0.00	100.00

C.2 Road way roughness data along Morogoro road

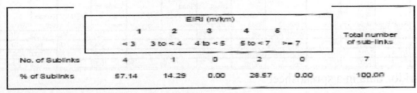

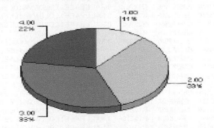

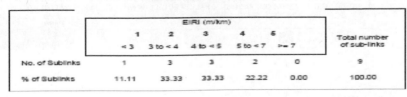

	EIRI (m/km)					
	1	**2**	**3**	**4**	**5**	Total number of sub-links
	< 3	3 to < 4	4 to < 5	5 to < 7	>= 7	
No. of Sublinks	1	3	3	2	0	9
% of Sublinks	11.11	33.33	33.33	22.22	0.00	100.00

D.Proximity data

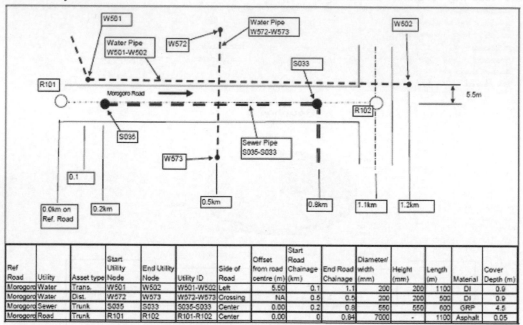

Ref Road	Utility	Asset type	Start Utility Node	End Utility Node	Utility ID	Side of Road	Offset from road centre (m)	Start Road Chainage (km)	End Road Chainage	Diameter/ width (mm)	Height (mm)	Length (m)	Material	Cover Depth (m)
Morogoro	Water	Trans.	W501	W502	W501-W502	Left	5.50	0.1	1.1	200	200	1100	DI	0.9
Morogoro	Water	Dist.	W572	W573	W572-W573	Crossing	NA	0.5	0.5	200	200	500	DI	0.9
Morogoro	Sewer	Trunk	S035	S033	S035-S033	Center	0.00	0.2	0.8	550	550	600	GRP	4.5
Morogoro	Road	Trunk	R101	R102	R101-R102	Center	0.00	0	0.84	7000	-	1100	Asphalt	0.05

E. Road corridor data in a spreadsheet

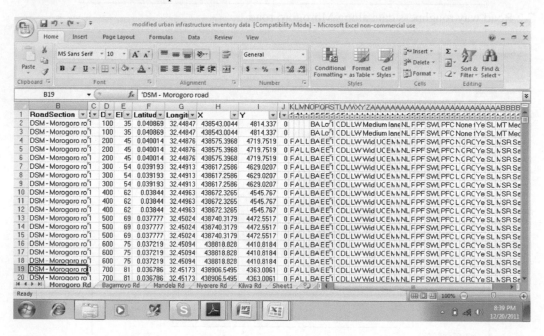

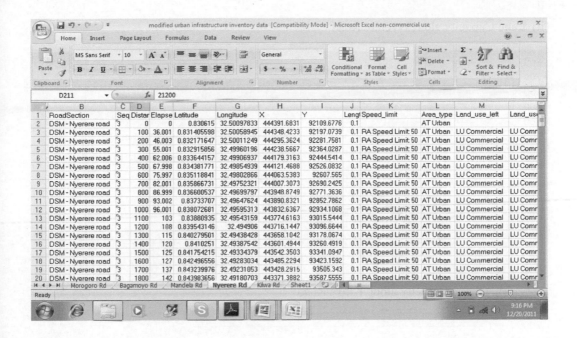

F. Lane capacity
(i) Lane Capacity along Morogoro Road

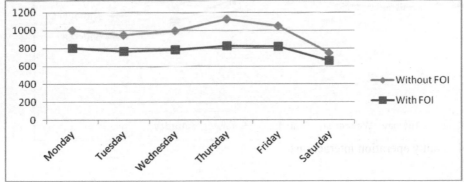

(ii) Lane Capacity along Nyerere Road

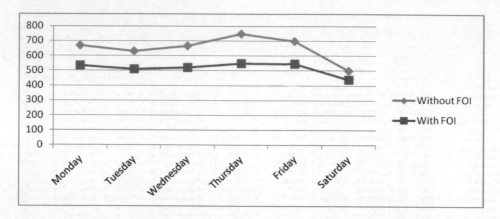

(iii) Lane Capacity along Kilwa Road

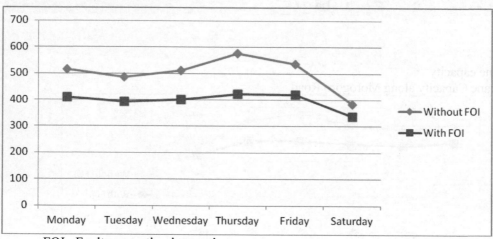

FOI –Faulty operation interactions

G. Travel time during the congestion

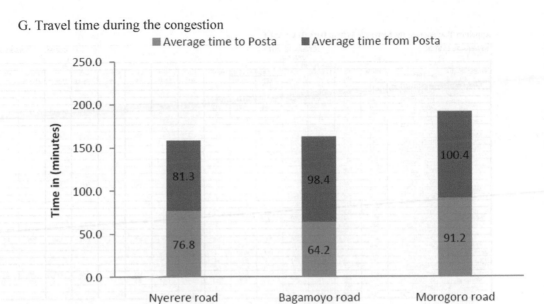

■ Average time to Posta   ■ Average time from Posta

Appendix B.1: Some of the DAWASCO Water Supply Pipes Data

Object Id	Id	Ref	Comission Date	Project Status Name	Material	Nominal Diameter (mm)	Thickness (mm)	Chain Id	No of Connections	No of Valves	User Length	GIS Length
1	1	2551	19/09/2006	Existing	MS	525	10	5354	0	0	315.49	315.53
2	2	Water Pipe 2553	27/09/2006	Existing	Steel - B	525	4.5	1	0	0	46.28	46.29
3	3	Water Pipe 57	18/05/2006	Existing	Cast Iron	150	12	974	0	1	84.69	84.65
4	4	Water Pipe 60	18/05/2006	Existing	Cast Iron	150	12	975	0	0	269.9	269.9
5	5	Water Pipe 61	18/05/2006	Existing	Cast Iron	150	12	981	0	0	52.24	52.25
0	6	Water Pipe 62	18/05/2006	Existing	Cast Iron	75	12	982	0	0	151.41	151.42
7	7	Water Pipe 64	18/05/2006	Existing	Cast Iron	75	12	983	0	0	99.79	99.8
8	8	Water Pipe 65	18/05/2006	Existing	Cast Iron	75	12	986	0	0	67.9	67.91
9	9	Water Pipe 66	18/05/2006	Existing	Cast Iron	75	12	980	0	0	105.44	105.45
10	10	Water Pipe 67	18/05/2006	Existing	Cast Iron	75	12	987	0	0	64.69	64.69
11	11	Water Pipe 70	18/05/2006	Existing	Cast Iron	75	12	988, 990, 998, 1000	0	1	643.55	643.51
12	12	Water Pipe 71	18/05/2006	Existing	Cast Iron	75	12	999	0	0	187.53	187.54
13	13	Water Pipe 72	18/05/2006	Existing	Cast Iron	75	12	992	0	1	169.57	169.59
14	14	Water Pipe 73	18/05/2006	Existing	Cast Iron	200	12	991	0	1	80.9	80.91
15	15	Water Pipe 74	18/05/2006	Existing	Cast Iron	200	12	993	0	0	107.82	107.84
16	16	Water Pipe 75	18/05/2006	Existing	Cast Iron	75	12	976, 977, 979, 959,	0	1	776.15	776.27
17	17	Water Pipe 76	18/05/2006	Existing	Cast Iron	150	12	996, 965, 963, 961,	0	0	639.29	639.33
18	18	Water Pipe 77	18/05/2006	Existing	Cast Iron	100	12	928	0	1	81.57	81.57
19	19	Water Pipe 78	18/05/2006	Existing	Cast Iron	75	12	962	0	0	193.44	193.47
20	20	Water Pipe 79	18/05/2006	Existing	Cast Iron	150	12	995, 968	0	0	204.78	204.82
21	21	Water Pipe 80	18/05/2006	Existing	Cast Iron	75	12	994	0	1	224.75	224.78
22	22	Water Pipe 2571	28/09/2006	Existing	PCP	1350	13.5	6359	0	0	1.37	1.37
23	23	Water Pipe 95	18/05/2006	Existing	Cast Iron	375	10	1281	0	0	373.64	373.67
24	24	Water Pipe 96	18/05/2006	Existing	Cast Iron	375	10	1215, 1248, 1249, 1	0	0	1300.28	1300.39
25	25	Water Pipe 97	18/05/2006	Existing	Cast Iron	375	10	1241, 1282, 1242, 1	0	0	737.1	737.16
26	26	Water Pipe 98	18/05/2006	Existing	DI	395	10	1244	0	0	163.73	163.74
27	27	Water Pipe 2555	23/05/2006	Existing	PCP	1350	13.5	5733	0	0	1266.95	145.08
28	28	Water Pipe 101	18/05/2006	Existing		0	0	997	0	0	258.41	258.43
29	29	Water Pipe 102	18/05/2006	Existing	Cast Iron	150	12	1009, 1010, 1011, 1	0	1	1412.39	1412.56
30	30	Water Pipe 103	18/05/2006	Existing	Cast Iron	75	12	1002, 1003	0	1	315.93	315.98

Appendix B.3: Some of the Sewerage System Data (DAWASCO)

STREET/ROAD	Asset Reference	FROM MH	TO MH	LENGTH (m)	DIA (mm)	INVERT LEVEL Upper	INVERT LEVEL Lower	SLOPE	Qfull (l/s)	Vfull (m/s)	Land Use	Houses nos	Avg Flow l/s	Peak Flow l/s	Proport Flow %	Velocity m/s	ANALYSIS Remarks
Kalenga st.	MH741-MH742	741	742	75	250	13.11		6									
	MH742-MH743	742	743	75	250			0									
	MH743-MH744	743	744	75	250		11.44	0									
	MH744-MH745	744	745	91	250	11.44	11.01	212	32.56	0.66							
	MH745-MH746	745	746					0									
Kalenga st.	MH746-MH747	746	747	77	250	10.88	10.36	241	30.78	0.63							
	MH747-MH748	747	748					0									
Kalenga st.	MH748-MH749	748	749	73	250	10.27	9.95	228	31.51	0.64	MD	68	0.85	1.95	17%	0.35	
Malik road	MH749-MH740	749	740		250	9.95											
Malik road	MH740-MH738	740	738		250			0									
Mindu st.	MH732-MH733	732	733					0									
	MH733-MH734	733	734					0									
	MH734-MH735	734	735					0									
	MH735-MH736	735	736					0									
	MH736-MH738	736	738					0			MD	68	0.65	1.95			
Malik road	MH738-MH731	738	731					0									
	MH731-MH730	731	730		250												
	MH730-MH717	730	717		250	7.00	6.83										
	MH717-MH703	717	703	98	300	6.83	5.99	117	72.05	1.02							
Mazengo Rd.	MH673-MH672	673	672	60	200	9	8.77	261	16.01	0.51							
	MH672-MH671	672	671					0									
	MH671-MH670	671	670					0									
	MH670-MH669	670	669	81	200	8.4	8.1	270	15.8	0.5	MD	54	0.52	1.56	22%	0.31	
Kibasila st.	MH669-MH686	669	686	77	450	7.81	6.95	90	243.83	1.53							
Mathuradas Rd.	MH690-MH689	690	689	90	200	12.1	10.07	44	39.45	1.26							
Mathuradas Rd.	MH689-MH688	689	688	54	200	10.07	8.79	42	39.45	1.26							
Mathuradas Rd.	MH688-MH687	688	687	73	200	8.79	8.22	128	22.95	0.73							
Mathuradas Rd.	MH687-MH686	687	686	89	200	8.22	6.95	70	30.77	0.98	MD	56	0.54	1.62	16%	0.5	
Kibasila st.	MH686-MH705	686	705	77	250	6.95	6.2	103	46.81	0.95	MD		1.06	3.18	18%	0.53	
Sevya St.	MH708-MH707	708	707					0									
	MH707-MH706	707	706	100	150	9.3	6.95	43	18.26	1.03							

Annex III

A. Infrastructure installation and operations guide

1.0   INTRODUCTION

1.1 Scope

This guide stipulates the required mechanisms for preparation and management of applications, issuance and monitoring of compliance to the permit terms to be issued by road authority to utility services providers for installation and repair of their infrastructures within the road corridor. Such services include, but are not limited to electricity transmission lines, gas pipes, oil pipes and any other energy supply lines and installations, telephone lines, mobile phone cables and towers, optic fibre cables, urban railway lines, water pipes, sewerage pipe networks and any other services that will be laid over, across, on or beneath the ground surface within the road corridors.

1.2 Objectives

This Guide aims at establishing a monitoring system in order to:

- Mitigate uncalled for encroachments and subsequent likely future compensation costs that arise during new road construction or reconstruction, when relocation of infrastructure is necessary;
- Economical use of the limited space by provision of a duct where possible that shall be shared by various utility service providers;
- Encourage coordination among the infrastructure provider
- Contribute to improved movement and safety;
- Encourage the use of shared duct as opposed to non-ducted underground infrastructure.

1.3   Advantages of the Guide

The Guide will realize the following benefits:

- Guide the road authority practices in processing and approving/ disapproving applications for infrastructure installation and repair;
- Facilitate pre-planning of activities related to issuance and supervision of compliance to temporary permits issued by road authority;
- Form a basis for prospective applicants in planning their infrastructure installation strategies;
- Provide reference for infrastructure service providers in preparing applications;

- Become a precedent document for publicly acceptable management of infrastructure provision within and, where warranted, outside the road corridor, which may also be adopted by other organisations; and,
- Encourage standardization of assessment criteria and uniformity in the application of these criteria.
- Provide shared data base of the infrastructure network layout
- Facilitate Integration of urban infrastructure
- Encourage coordination of urban infrastructure operators

2.0   PROCEDURES
2.1   Mode of Application

All applications of permits for installation or operations of the infrastructure in road corridor will be done in the following manner:-

2.1.1   Any request may be done by an individual person, a company or a corporation as the case may be, who intends to operate for the purpose of its own service provision or by agents.

2.1.2   The application for such a permit shall be in writing to the road authority.

2.1.3   Such application shall include full particulars of the applicant, the proposed   descriptions/specifications,   and   appropriate opportunities/locations, drawings showing key dimensions (layout plan, typical cross sections, and parts details), necessary calculations and design details as well as photographs as relevant.

2.1.4   Upon receipt of the application, the road authority will invite the applicant, specifying the date for joint site inspection and appoint relevant road authority official to supervise the process of such inspection.

2.1.5   The road authority official so appointed shall be accompanied by the representative of the applicant to visit and inspect the location.

2.2 Site Visits and Inspection Form

2.2.1   At the site, the assigned road authority official and the applicant's representative will jointly inspect the location and discuss its suitability of accommodating the services on the basis of this Guide and fill in appropriate **Inspection Form No.1**

2.2.2   Both parties shall sign the form to confirm the record of their inspection.

2.2.3   During the site inspection, the decision on choosing a location to new utility   shall consider possibility of separating overhead infrastructure with underground infrastructure on opposite sides of the road corridor.

2.3 Evaluation of Applications

2.3.1 Upon completion of a joint inspection, the road authority official shall evaluate the application based on the information presented in the drawings and any other submission in the application in comparison to the facts collected at site, assessing as to whether they comply to the Guide requirements and by ticking in appropriate spaces.

2.4 Approval or Refusal of Permit

2.4.1 Immediately after completion of inspection report and ensuring that all requirements for a particular application have been observed, then approval or refusal may be given within fourteen days.

2.4.2 Once the application and permit presented in a format shown in **Form 2**, the applicant shall be informed in writing.

2.4.3 The permit shall be issued and become effective.

2.5 Compliance to the permit

2.5.1 The road authority shall ensure that the installation of the approved utility is done on site in accordance with the approved position in terms of offset from the road centreline, type, shape, size, height, width, depth, clearance, and that the applicant complies with the appropriate specifications.

2.5.2 Soon after installation of the utility is complete, the service provider or investor shall ensure compliance to any other terms of the permit such as providing and fixing of surface markers to locate an underground installed utility and making necessary public awareness on the presence of the utility, which shall be coordinated under the close supervision of the road authority.

2.5.3 The service provider or investor shall indemnify the road authority from any third party claims and those of their staff with respect to injury and death to persons as well as any damage to property. The road authority shall further be indemnified from any claims that may arise from the effects that may arise as a result of installing the utility, including any other fees or charges payable to other parties and authorities.

3.0 CONDITIONS

The infrastructure shall be categorised into three major groups, namely:
i. Overhead Infrastructure.
ii. Ducted underground Infrastructure.
iii. Trenched or Non-ducted Underground Infrastructure.

The permits for utility and other related customers shall bear the following conditions;

3.1 Installation and ownership

Overhead, Ducted underground and Trenched or Non-ducted Underground Infrastructure shall be installed and owned by respective investor or service provider.

3.1 Sharing of Road Corridor

3.1.1 Duct Sharing

The sharing of duct is encouraged among all communication and other service providers or investors as a way of increasing the choice to end users, and helps foster the roll out of new competitive services. It shall be designed and constructed so as to reduce the capital cost to the parties and impact on the environment, and enable an economic use of the limited road corridor.

3.1.2 Overhead and Trenched or Non-ducted Underground Infrastructure
The permitted owners or investors of these infrastructures are encouraged to share the same road corridor but in a separate corridor located and specified in the permit issued to them by road authority.

3.3 Management and maintenance of utility

3.3.1 The utility infrastructures shall be managed and maintained by the respective owners.

3.3.2 In case the utility owner requires opening up the land in the road reserve for purposes of repair or maintenance of underground utility, it shall save at least three days prior notice to the road authority.

3.3.3 The utility service provider or investor shall maintain all information related to the locations of the utility structures and be responsible to identify them by fixing marker posts for purpose of mitigating damages during road reconstruction or maintenance of the other infrastructure in the road corridors.

3.4. Specifications for installation of the utility

3.4.1 The utility infrastructures shall be installed at the exact location described in the permit while observing all requirements of road safety, positioning, spacing, design, construction, maintenance and traffic movement.

3.4.2   The size, area, clear depth or height of the utility infrastructures shall be observed and their maximum or minimum measures shall not exceed the limit indicated in the permit.

3.4.3   Where the utility runs along any road, the utility shall be placed at the extreme end of road reserve as far as practicable at the depth where the earth cover shall not be less than 1.5 metres.

3.4.4   Where the utility crosses beneath any paved road, there shall be HDD boring/tunnelling at the depth where the earth cover shall not be less than 1.5 meters below the road level and shall install the utility with a length extending to cover the existing drainage systems; except where open trenching is unavoidable, relatively larger infrastructures are to be installed and followed by acceptable restoration of the road to meet the required original standards.

3.4.5   Where the utility crosses beneath any unpaved road, there shall be trenching at the depth where the earth cover shall not be less than 1.5 meters below the road level, and thereafter a backfilling and compaction to meet acceptable engineering standards to restore the road. In case the utility crosses the inlet/outlet of the road-drainage structure or mitre drain, it shall be laid below the invert level of the drainage channel by a minimum of 300mm and protected by concrete structure of a minimum thickness of 200mm, class 20 concrete.

3.4.6   Where the utility crosses or run along any bridge, there shall be made an independent supporting structure for the utility.

3.4.7   The utility investor or owner shall install marker posts along the utility line at the interval of one hundred meters (100m) within urban areas

3.4.8   Upon completion of the installation of the utility, the owner or an investor shall submit As-Built information of survey data to road authority. The data of points shall comprise of Universal Traverse Mercator (UTM) coordinate system of the utility location at minimum interval of fifty (50) meters. The UTM coordinated points shall define positions of utility in terms of Northing, Easting and Datum level above mean sea level tied to the National Grid.

FORM 1:

SITE INSPECTION FORM FOR UTILITY INSTALLATION
(To be filled and attached to the approved Permit)

SERIAL No. ....

1.   NAME OF APPLICANT: ....
2.   FULL PERMANENT ADDRESS: ....
3.   REGISTRATION No. OF APPLICATION: .... DATE: ....
4.   LETTER OF SITE INSPECTION: REF. No: ....DATE: ....
5.   ROAD NAME APPLIED: ....
6.   LOCATION/CHAINAGE* FROM: .... TO: ....
7.   LOCATION SKETCH (See attachment)
8.   DETAILED PARTICULARS FOR THE UTILITY:

POSITION: Underground/Surface/Overhead*

LAYOUT: Along/Across the road/across* a bridge ....

DEPTH (Below ground)/HEIGHT (Above ground)*:........................(m)

EXCAVATION METHOD: Open Trenching/Boring(HDD)

SLEEVE/SERVICE CONDUIT: Required/Not required

DISTANCE FROM ROAD CORRIDOR END: ....(m)

ADDITIONAL COMMENTS: ....

Prepared on behalf of road authority by:

.... (Name)

Signature.................Title..................... Date ....

Accepted on behalf of Applicant by:

.... (Name)

Signature.................Title..................... Date ....

9.   **EVALUATION OF SITE INFORMATION AND RECOMMENDATION:**
I .... (Name) certify that upon checking and evaluation of site information, I DO HEREBY **RECOMMEND/NOT RECOMMEND*** this application for utility because of...
 (Signature) ....Title .... Date ....

10.   **APPROVAL BY AUTHORISED OFFICER**:
I .... (Name) have reviewed the recommendation made for utility installation and I THEREFORE **APPROVE /NOT APPROVE*** the permit because of ....
Signature ....Title...................Date ....

   - *Delete whichever is not applicable.*

FORM 2:

INFRASTRUCTURE INSTALLATION PERMIT

PERMIT NO: .... FY ....

1. NAME OF THE INVESTOR/SERVICE PROVIDER*: ....

2. FULL ADDRESS OF APPLICANT: ....

3. RECEIPT NO: ....

4. ISSUING OFFICE: ....

5. UTILITY SIZE: ....

6. DESCRIPTION OF THE INFRASTRUCTURE ALONG/ACROSS*;
   ROAD NAME..
   AREA/LOCATION*: ....
   EARTHCOVER /OVERHEAD CLEARANCE (m): ....

7. PERMIT VALIDITY PERIOD: ....
   EFFECTIVELY FROM ....
8. EXPIRY DATE: ....

9. LOCATION SKETCH: (See separate sheet)

Dated at ....this......... day of ....month ....of 20.........

Issued by:

   Name ....
   Signature ....
                    ....
                    Road Authority

B. Model results
B.1 Road corridor safety feature index along Morogoro and Bagamoyo roads

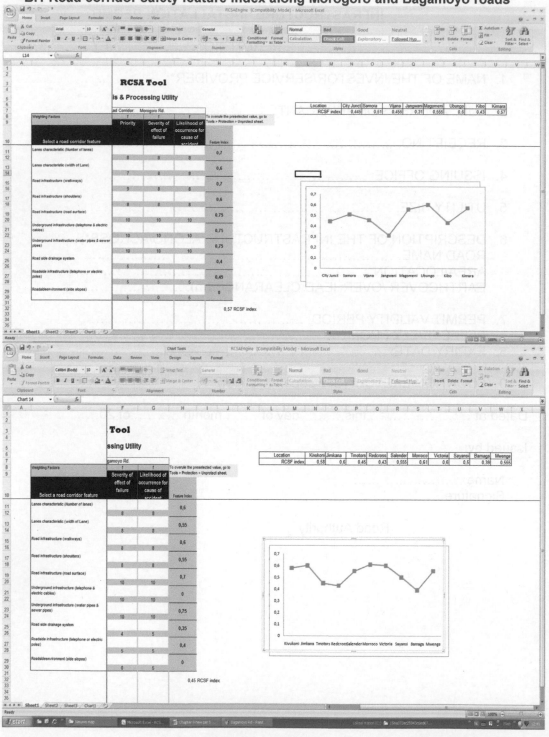

B.3 Infrastructure Interactions versus Road Conditions

When Infrastructure interactions are regressed against the road conditions properties it is observed that there is a negative relationship between interactions and road conditions as signified by the Model fit ANOVA where the p-value < 0.01. This means when there are so many interactions in the road corridor the road the road safety becomes worse. This is vivid whereby there is a negative relationship by R=-0.832. The Adjusted R is 68.8 % means that 68.8% of the road conditions are determined by the interactions. Moreover the coefficients have proved to be significant and these results are presented in the tables below.

Model Summary

Model	R	R Square	Adjusted R Square	Std. Error of the Estimate
1	-.832[a]	.691	.688	.279

a. Predictors: (Constant), Functional Interactions, Urban Physical interactions, Road surface Interactions

ANOVA[b]

Model		Sum of Squares	Df	Mean Square	F	Sig.
1	Regression	42.993	3	14.331	183.780	.000[a]
	Residual	19.183	246	.070		
	Total	62.176	249			

a. Predictors: (Constant), Functional Interactions, Urban Pysical interactions, Road surface Interactions

b. Dependent Variable: Road Conditions

Coefficients[a]

Model		Unstandardized Coefficients		Standardized Coefficients	t	Sig.
		B	Std. Error	Beta		
1	(Constant)	.178	.058		3.088	.002
	Road surface Interactions	.261	.068	.258	3.816	.000
	Urban Pysical interactions	.402	.065	.402	6.172	.000
	Functional Interactions	.230	.075	.228	3.049	.003

a. Dependent Variable: Road Conditions

B.4 Road condition and safety of movement

ANOVA[b]

Model		Sum of Squares	df	Mean Square	F	Sig.
1	Regression	31.843	1	31.843	264.361	.000[a]
	Residual	29.873	248	.120		
	Total	61.716	249			

a. Predictors: (Constant), Road Conditions

b. Dependent Variable: Safety of Movement

Model Summary

Model	R	R Square	Adjusted R Square	Std. Error of the Estimate
1	.718[a]	.516	.514	.347

a. Predictors: (Constant), Road Conditions

Coefficients[a]

Model		Unstandardized Coefficients		Standardized Coefficients	t	Sig.
		B	Std. Error	Beta		
1	(Constant)	.396	.068		5.822	.000
	Road Conditions	.716	.044	.718	16.259	.000

a. Dependent Variable: Safety Movement

C. Proposed road corridors

C.1 Proposed 4 lanes 2 ways with underground infrastructure placed under the walkways

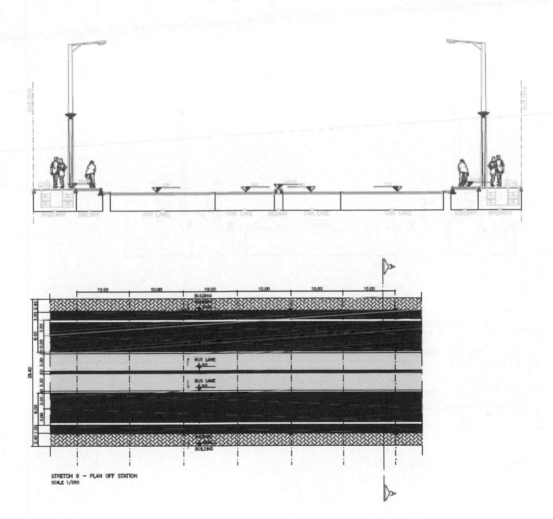

STRETCH 9 — PLAN OFF STATION
SCALE 1/250

C.2 Proposed 2 lanes 2 ways with underground infrastructure placed under the walkways

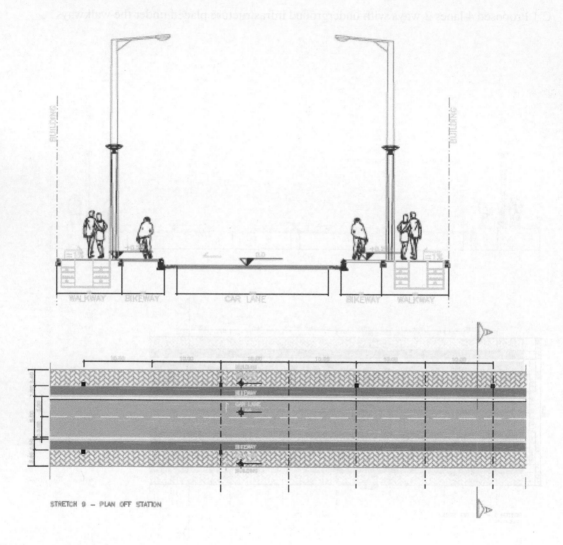

STRETCH 9 — PLAN OFF STATION

Annex IV

A. Laying and repair of underground water pipes

B. Cut of road for underground repair and not properly reinstated

C. water leakages and congestion

C. Excavation for sewer pipe repair

Labourers dig trenches to replace worn out sewage pipes on Sewa Street in Dar es Salaam yesterday. Rehabilitation of broken sewers gives hope that things might be different during this rainy season when flood waters wreak havoc in the city. (*Photo: Moshy Kiyungi*)

T - #0083 - 071024 - C98 - 240/170/14 - PB - 9780415627146 - Gloss Lamination